This series fosters information exchange and discussion on all aspects of manufacturing and surface engineering for modern industry. This series focuses on manufacturing with emphasis in machining and forming technologies, including traditional machining (turning, milling, drilling, etc.), non-traditional machining (EDM, USM, LAM, etc.), abrasive machining, hard part machining, high speed machining, high efficiency machining, micromachining, internet-based machining, metal casting, joining, powder metallurgy, extrusion, forging, rolling, drawing, sheet metal forming, microforming, hydroforming, thermoforming, incremental forming, plastics/composites processing, ceramic processing, hybrid processes (thermal, plasma, chemical and electrical energy assisted methods), etc. The manufacturability of all materials will be considered, including metals, polymers, ceramics, composites, biomaterials, nanomaterials, etc. The series covers the full range of surface engineering aspects such as surface metrology, surface integrity, contact mechanics, friction and wear, lubrication and lubricants, coatings an surface treatments, multiscale tribology including biomedical systems and manufacturing processes. Moreover, the series covers the computational methods and optimization techniques applied in manufacturing and surface engineering. Contributions to this book series are welcome on all subjects of manufacturing and surface engineering. Especially welcome are books that pioneer new research directions, raise new questions and new possibilities, or examine old problems from a new angle. To submit a proposal or request further information, please contact Dr. Mayra Castro, Publishing Editor Applied Sciences, via mayra.castro@springer.com or Professor J. Paulo Davim, Book Series Editor, via pdavim@ua.pt

More information about this subseries at http://www.springer.com/series/10623

Jagadish · Kapil Gupta

Abrasive Water Jet Machining of Engineering Materials

 Springer

Jagadish
Department of Mechanical Engineering
National Institute of Technology
Raipur, Chhattisgarh, India

Kapil Gupta
Department of Mechanical and Industrial
Engineering Technology
University of Johannesburg
Johannesburg, South Africa

ISSN 2191-530X   ISSN 2191-5318   (electronic)
SpringerBriefs in Applied Sciences and Technology
ISSN 2365-8223   ISSN 2365-8231   (electronic)
Manufacturing and Surface Engineering
ISBN 978-3-030-36000-9   ISBN 978-3-030-36001-6   (eBook)
https://doi.org/10.1007/978-3-030-36001-6

This Springer imprint is published by the registered company Springer Nature Switzerland AG
The registered company address is: Gewerbestrasse 11, 6330 Cham, Switzerland

Preface

Advanced machining processes have been explored as a viable alternate to conventional machining methods. Recently, some efforts have also been done to make these processes green or sustainable. Abrasive water jet machining process is one of the most important advanced machining processes and also recognized as a green process. The main objective of this book is to present the capability of abrasive water jet machining process to machine a wide range of engineering materials and to facilitate specialists, engineers, and scientists to establish the field further. This book consists of four chapters. It starts with Chap. 1 as an introduction to abrasive water jet machining process where its working principle, advantages, limitations, applications, and literatures are discussed. Chapter 2 presents aspects of machining metallic materials by abrasive water jet process. Abrasive water jet machining of polymer (wood dust filler-based reinforced) composites is reported in Chap. 3. It also presents the optimization of abrasive water jet machining process by MOORA technique to secure the enhanced machinability of polymer composites. The last chapter Chap. 4 is focused on experimental investigation and process optimization for machining of zirconia ceramic composites by abrasive water jet process.

The information presented and investigation results reported in this book are from the research conducted by the authors in this area. Authors hope that the research reported on the experimentation, modeling, and optimization would facilitate and motivate the researchers, engineers, and specialists working in this area.

We sincerely acknowledge Springer for this opportunity and their professional support.

Raipur, India   Jagadish
Johannesburg, South Africa   Kapil Gupta

v

Contents

Chapter 1
Introduction to Abrasive Water Jet Machining

1.1 History and Background

The cutting technology by a jet of water (high-pressure water erosion) was firstly introduced in the middle of 1800s to cut rocks and for mining applications [1]. Many years later, sometimes around 1950, this technology was used for cutting soft materials like paper. In the 1980s, abrasives as media were introduced in water jet to enhance the process efficiency. Motion control systems and process flexibility were the major developments as regards to this technology during 1990. Since then to now, a series of developments to accomplish machining hard and brittle material, manufacturing typical shapes and micro-products, cleaning and polishing, and for developing biomedical, scientific, and electronic components. Electrochemical slurry jet machining, AWJM with ice particles as media, innovations in nozzle design, mixing polymer additives in abrasives, and process parameter optimization, etc., are some of the major aspects of research, developments, and innovations in this technology [2].

1.2 AWJM Working Principle and Process Parameters

Abrasive water jet machining is an extended version of water jet machining where abrasive particles such as aluminum oxide, silicon carbide, or garnet are contained within the water jet with the purpose of raising the rate of material removal beyond that of a water jet machine [3, 4]. Abrasive water jet machining process can be employed to a wide range of materials that are soft from rubbers and foam to hard brittles ones like metals, ceramics, and glass. With movements that are computer driven, the cutting stream is therefore allowed to make objects efficiently and accurately. Materials that are difficult to cut through thermal cutting or by laser cutting can ideally cut through the AWJM process. Figure 1.1 illustrates the schematic of

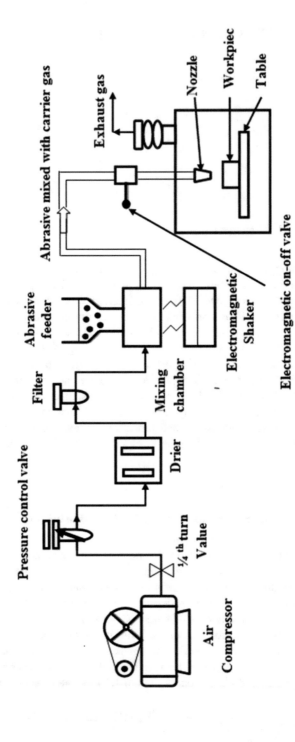

Fig. 1.1 Schematic diagram of abrasive water jet machining setup [4]

AWJM system. A typical abrasive water jet system consists of a pump system of high pressure, a catcher unit, abrasive feed system, a position control system, an abrasive water jet cutting head, and a water supply system.

This manufacturing technique operates on the mechanical erosion principle in which acceleration of abrasive slurry through a high-pressure fine jet is used to cut the material of interest [3, 4]. Water jet velocity together with an abrasive is on average equivalent to 300 m/s, while in other special applications the water jet and abrasive velocity go up to 900 m/s. High kinetic energy of the mixture results from the high velocity leading to rapid erosion of the workpiece targeted.

An abrasive water jet machine normally has an operating pressure of roughly 300–450 MPa, and this is sufficient to result in water jet velocity as high as 900 m/s. A rise in kinetic energy of water jet mixture is due to an increase in water pressure, leading to increased momentum that transfers to the abrasive particles, and the target material is cut by the impact and momentum changes of the abrasive material [3, 4]. The high-velocity abrasive water jet possesses the ability to cut different types of materials including metals, composites, ceramics, and rocks.

The removal of material occurs through erosion wear on the upper surface and is then followed at lower regions of workpiece being cut by wear deformation [3–5]. The schematic diagram in Fig. 1.2 illustrates the basic working principle with a cutting head assembly of the AWJM process.

The operating principle of the abrasive water jet system is such that water is pumped to high pressure by a high-pressure system that delivers it through the entire system. The hydraulic oil is pressurized by a hydraulic radial displacement pump and is then fed to an intensifier pump driven by the hydraulic oil, giving a carrier medium for the abrasive material. The water pump pressure ranges from 150 to 450 MPa. Transporting water from the pump at high pressure to the cutting head is done by using the water supply system. Different types of valves and joints together with high-pressure pipes make up the major elements of the system in supplying water. Water flow rate can go up to 11 l in a minute. The abrasive feed system is responsible for mixing water and the abrasive particles.

The abrasive feed system is made up of a delivery hose, a metering valve, and an abrasive hopper. Abrasive particles are transported to the abrasive inlet on the nozzle assembly from the metering valve through the delivery hose. An ON/OFF switch is the function of the metering valve by controlling the abrasive mass flow rate. A container used to store abrasives is considered to be the hopper. To achieve optimal cutting efficiency, it is necessary that the distance between the abrasive hopper and nozzle assembly must be minimal.

The cutting head otherwise known as the nozzle assembly is made up of the following components mixing chamber, water jet nozzle, and an orifice. The nozzle assembly otherwise known as the cutting head consists of a mixing chamber, an orifice, and an abrasive water jet nozzle. The cutting head is for providing and controlling the high-pressure water jet. It consists of diameters typically ranging between 1 and 5 mm, and the orifice is responsible for transporting the high-pressure water into high-velocity water jet. Diamond and sapphire are some most commonly used

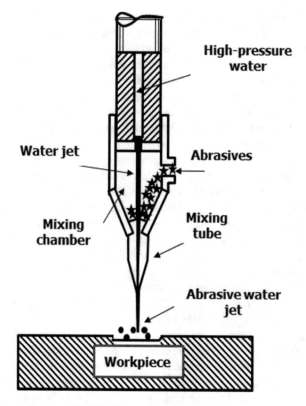

High-pressure water

Water jet

Abrasives

Mixing chamber

Mixing tube

Abrasive water jet

Workpiece

Fig. 1.2 Working principle of AWJM [1]

materials for an orifice [4, 5]. A diamond orifice typically lasts up to 2000 h, while the sapphire orifice can only work for 200 h. A space where the abrasive particles and water are mixed and entrained by vacuum suction is where the mixing chamber is placed in between the nozzle and orifice.

Preceding the mixture of the water jet within the mixing chamber is the acceleration of abrasive particles through the nozzle. The basic material of the nozzle is tungsten carbide. One way to obtain cutting performance that is optimal is through inflexible requirements for nozzle length and diameter. If the nozzle diameter is predetermined on the basis of the orifice diameter, this then dictates that would roughly be 2.5 up to 5 times the size of the diameter of the orifice. It is important that the nozzle length is accurate because a shorter than required nozzle length prohibits the assurance of required acceleration of the abrasive particles, whereas a longer than required one can lead to excessive wear rate due to passing abrasive water jet, which will also cause a loss in momentum.

The principle of mechanical erosion is a mechanism of which abrasive water jet machining uses to remove or cut material from a workpiece. This principle is based on abrasive particles mixed with a high stream of water jet velocity to target a given specimen [4, 5]. The specimen is under high stress due to kinetic energy that converts into pressure energy, and particles from the target material are inevitably removed as the induced stress reaches a peak point.

There are several important processing parameters of abrasive water jet machining that contribute to the effectiveness, performance, and efficiency of the AWJM process. The important AWJM parameters are as follows [1, 5]:

- **Abrasive Type**: In the process of abrasive water jet machining, the type of abrasive used is found to be the most important parameter as it is directly linked to the rate of material removal and the accuracy of machining. Hardness is normally the basis of selection when it comes to choosing the abrasive type; as such, the harder the material the harder the abrasive particle. The machining accuracy and rate of material removal are also critical parameters of the abrasive type. In general, sodium bicarbonate, glass beads, garnet, silicon carbide, aluminum oxide, silica, glass beads, and crushed glass are some of the most widely used abrasive types in the abrasive water jet machining process. For polishing, cleaning, and etching operations, dolomites and glass beads are most commonly used. Surface roughness can significantly be reduced through the use of harder abrasive materials, while on the other hand it also increases the rate of material removal of which subsequently reduces the process time. A few of the widely used abrasive types are illustrated in Fig. 1.3.
- **Abrasive Grain Size**: In abrasive water jet machining process, a significant role is played by the size of the abrasive particles or grain size. Fine particles are generally used for polishing and finishing operations, while the coarse grain particles are normally used for the cutting process. Moreover, smaller mesh number abrasives tend to have greater size in the average value of the particle and much less particles per unit weight. Figure 1.4 represents fine and coarse grain particle sizes used in AWJM.
- **Water Jet Pressure**: Flow of the abrasive depends on the pressure of water in order to cut the workpiece in AWJM. Kinetic energy of abrasive particles tends to rise with the increase in pressure of water and consequently raising the particles'

     (a)   Garnet   (b) Silicon Carbide   (c) Aluminium Oxide

Fig. 1.3 Types of abrasives

(a) Coarse Garnet   (b) Fine Garnet

Fig. 1.4 Coarse and fine abrasives

ability to removal material while decreasing the surface roughness, and the overall process duration is lowered. Surface quality is therefore improved as a result of higher water pressure as it drives up particle velocity and fragmentation within the abrasive nozzle. However, if the pressure of water is too high, it may generate a negative impact on the material. Extreme pressure of water leads to abrasive particles losing their ability to cut when they become too fragmented. In effect, a very high value of water pressure boosts the material removal rate (MRR), reduces the process time and yet increases surface roughness. Particle velocity is determined by the process through the water jet pressure; 450 MPa is the allowable high pressure; however, the normal range is between 100 and 300 MPa.

- **Standoff Distance**: One other significant parameter in abrasive water jet machining process is the standoff distance (SoD). The standoff distance is the space existing from the work surface to the tip of the nozzle. A larger standoff distance gives the jet the opportunity to expand prior to impinge the work surface. This larger standoff distance results in an increase in jet diameter during the cutting process which then leads to a reduction in the kinetic energy of the jet at impingement and therefore inefficient cutting and poor part quality. The deformation wear zone of material removal is influenced by the decrease in standoff distance and depth of cut. It is therefore more preferable to a low value of the standoff distance as it yields a smoother surface as a result of higher kinetic energy which finally increases the cutting process efficiency.
- **Abrasive Mass Flow Rate**: Cutting efficiency is directly linked to the abrasive mass flow rate in the abrasive water jet machining process. It is the rate that the abrasive flow of particles involved in the cutting and mixing process. The depth of cut also increases proportionally as a result of increased abrasive flow rate. The jet can easily cut through a workpiece and consequently enhancing cut surface smoothness and a greater rate of material removal when the flow rate of the abrasive is increased.

- **Traverse Speed**: The rate at which a nozzle head travels on the material being worked to achieve the process of machining. Less abrasive particles can attack the surface while increasing the rate of material removal, and a decrease in cut time results from high rate of traverse speed with reduced machining action. This ultimately degrades the quality of the surface. Thus, for good quality surface finish, it is important to lower the traverse rate.

1.3 Advantages, Limitations, and Applications of AWJM

The advantages offered by abrasive water jet machining technique are manifolds [1, 4]:

- It is an extremely versatile process suitable for a wide variety of materials.
- Unlike other advanced machining processes such as laser cutting and electric discharge machining-based processes, there are no heat-affected zones.
- Process offers high flexibility and independent of materials hardness and conductivity.
- The process needs low machining force.
- There is minimum material waste as a result of the cutting process.
- There is a lower likelihood of contamination of the environment.
- There is no requirement of lubricating or cooling oil in this process.
- During cutting, the process yields no toxic fumes and is thus regarded to be a green or an environmentally friendly process.

On the other hand, there are some limitations of this process, which are as follows:

- The capital cost of the process is quite high.
- There is high noise during the process.
- Some parts of the equipment have short life spans, i.e., orifice and the nozzle which increases the cost of overheads and cost of replacement to AWJM operation.
- Inaccurate combination of process parameters choosing can result in undesirable values of the geometry and surface roughness. Moreover, it leads to burrs that necessitate secondary finishing operations.

There are various potential applications of AWJM. A few industries can immediately benefit from AWJM technology due to the performance of the abrasive water jet economically and technically. Over the past few decades, abrasive water jet machining process was broadly utilized in different business sectors such as coal mining, manufacturing industry, civil and construction industry, food processing sector, cleaning and electronic business sectors. However, AWJM technology is mainly used in automotive, aerospace, mining, electronics and food industries, and some of the specific applications are discussed below [1, 4].

In the *automotive industry*, the technology is used to produce parts including but not limited to fiberglass body components, inside trimming like door panels, trunk liner, head liners and on the outside bumpers. Moreover, in *electronic industries*, AWJM is used to manufacture circuit boards and cable stripping.

Within the *aerospace business sector*, the technology is used to produce military aircraft body parts made from titanium, aluminum body parts, inner compartment parts plus engine components (stainless steel, titanium, aluminum, and heat-resistant alloys). AWJM technology has been primarily used for cutting materials regarded as "difficult to cut."

Abrasive water jet machining benefits the *coal mining commerce sector* through its ability to cut metals in a safe manner even in underground areas that may be potentially explosive.

In *construction industry*, abrasive water jet machining can be used in a number of useful applications including reinforced concrete cutting.

Within the *electronic commerce space*, the abrasive water jet cutting process is widely used for circuit boards cutting to make smaller pieces from a big stock. Water jet cutting of a minuscule kerf can easily be achieved with very little waste of materials. The water jet is able to cut to a tolerance near the required without causing any damage to parts that are fixed on the circuit board due to concentration.

An abrasive water jet machine can quite easily be fitted into a ship for offshore cutting system works in the *gas and oil* industry. Rescue operations, repairs, pipe cutting, platform cutting including oil well deactivation through cutting of casing some of the examples of use for AWJM with the industry of gas and oil.

The *food industry* may use this technology in preparation of food. The ice jet cutting process can easily be applied to trimming meat fats and the cutting of bread.

1.4 Past Work on AWJM of Engineering Materials

Nearly, over the last three–four decades, a considerable amount of research has been carried on and reported on AWJM of various engineering materials. Some of the work on AWJM of various engineering material are discussed here as under.

Babu and Muthukrishnan [6] conducted an investigation on AWJM of brass 360. Taguchi L_{27} orthogonal array-based experimental study where pump pressure, abrasive flow rate, standoff distance, and feed rate were found to significantly influence the surface finish. Optimal set of AWJM parameters, i.e., abrasive flow rate—75.37 g/min; pump pressure—399 MPa; standoff distance—1 mm; and feed rate—557 m/min, was obtained for optimum average roughness 5.19 μm. A study on surface roughness of aluminum alloy (AA 6351) when machining by AWJM was conducted by Babu et al. [7] where tremendous improvement in surface quality was achieved by increasing abrasive mass flow rate.

Wonder metal titanium and superalloy Inconel which are recognized as difficult-to-machine materials have also been machined by AWJ process. A detailed performance analysis of AWJM process parameters while machining Ti grade 5 was conducted by Vasanth et al. [8]. They found a strong relationship between abrasive mass flow rate and surface quality, especially roughness. Low mass flow rate with low standoff distance was recommended for better surface characteristics. During drilling and slotting of Ti-6Al-4V, traverse speed was identified as the most significant parameter affecting geometric characteristics of hole [9]. Similarly, blind pockets on Inconel superalloy for various industrial applications were fabricated by Bhandarkar et al. [10] using AWJM. Few available studies on machinability investigation and optimization on AWJM of Inconel identified standoff distance and water jet pressure as the parameters significantly affecting surface properties [11, 12].

Kevlar composite machining by AWJM was investigated by Siddiqui et al. [13] where 3.9 μm average roughness was achieved at water jet pressure—375 MPa, abrasive flow rate 300 g/min, and quality level 6. Another important research reports the successful machining of green composites by AWJM [14]. A detailed analysis, modeling, and optimization found the suitability of AWJM for quality machining of green composites.

Ruiz-Garcia et al. successfully cut quality straight cuts and holes in aluminum and composites [15]. Their investigation found the elimination of thermal damage and hence delamination of CFRP after machining by AWJM with high geometric accuracy of the parts. Jayakumar investigated the machinability of Kenaf/E-glass fiber-reinforced hybrid polymer composites under AWJM [16]. He obtained optimum set of AWJM parameters, i.e., water jet pressure: 255 MPa, abrasive flow rate: 0.275 kg/min or 4.6 g/s, SOD: 1.9 mm and traverse speed of 0.26 mm/min for the best values of average surface roughness—3.254 μm and kerf—0.255 mm. In an interesting study, Madhu and Balasubramanian engineered quality holes in CFRP by varying nozzle design in AWJM [17].

Xu and Wang studied the effect of nozzle oscillation on surface quality while machining alumina ceramic by AWJM [18]. Abrasive water jet cutting of silicon nitride ceramic is conducted by Ghosh et al. using silicon carbide abrasives [19]. The surface roughness was found to increase with increase in water jet pressure and decrease in abrasive mass flow rate. Some of the results of the investigation conducted by Unde et al. on AWJM of CFRP laminate reveal that laminate with low fiber orientation gives better results in terms of low kerf, high MRR, and better surface properties [20].

Various components such as micro-hard disk and memory card strips for electronics applications were successfully cut and fabricated by AWJM [21]. Pal and Choudhury used AWJM for fabrication of micro-pillars of good surface quality from different materials such as titanium alloys, aluminum, and stainless steel [22]. Pillars of different aspect ratios were achieved varying height in the range of 265–720 μm and taper ratio in the range of 10–15°.

Furthermore, abrasive water jet machining has also been used for polishing (of hard brittle materials) and cleaning applications [23]. Downsizing of AWJM's nozzle and orifice system efficiently helped for micromachining of precision thin structures and manufacture of micro-heat sinks [24]. The thickness of heat sinks or fins manufactured by AWJM varied from 150 to 700 μm. In an important recent attempt, micro-holes and micro-channels have successfully been fabricated in amorphous glass by AWJM machining technology [25].

Some literature also indicated about the green and sustainability aspects of abrasive water jet machining process [26–28]. For example, reduced wastage, no heat-affected zone, low environmental contamination, and no cutting fluid requirement make this process to be known as a sustainable process and can be further explored as a sustainable alternate to other conventional and advanced processes. An important recent study on the fabrication of miniature spur gears of brass by AWJM recognizes it as a sustainable process to produce quality products, at high productivity and with sustainability [29, 30]. However, sincere future attempts are required toward finding sustainable abrasives and their recycling, and lifecycle engineering and analysis of AWJM for machining of different engineered products from various materials.

Some experimental studies conducted by the authors on abrasive water jet machining of various engineering materials are discussed in the subsequent chapters of this book.

References

1. K. Gupta, M.K. Gupta, Developments in non-conventional machining for sustainable production—a state of art review. Proc. IMechE Part C J. Mech. Eng. Sci. (2019). https://doi.org/10.1177/0954406218811982
2. K. Gupta, M. Manjaiah, M. Avvari, A. Mashamba, Ice-jet machining—a sustainable variant of abrasive water jet machining, in *Sustainable Machining*, ed. by J.P. Davim (Springer International Publishing, 2017), pp. 67–78. ISBN 978-3-319-51961-6
3. M. Hashish, Cutting with abrasive-waterjets. Mech. Eng. **106**(3), 60 (1984)
4. S. Bhowmik, Jagadish, A. Ray, Abrasive water jet machining of composite materials, in *Advanced Manufacturing Technologies*, ed. by K. Gupta (Springer, 2017), pp. 77–97
5. V.K. Jain, *Advanced Machining Processes* (Allied Publishers, New Delhi, 2007)
6. M.N. Babu, N. Muthukrishnan, Investigation on surface roughness in abrasive water-jet machining by the response surface method. Mater. Manuf. Processes **29**, 1422–1428 (2014)
7. M.N. Babu, A.A. Fernando, N. Muthukrishnan, Analysis on surface roughness in abrasive water jet machining of aluminium. Prog. Ind. Ecol. **9**, 200–206 (2015)
8. S. Vasanth, T. Muthuramalingam, P. Vinothkumar, T. Geethapriyan, G. Murali, Performance analysis of process parameters on machining titanium (Ti-6Al-4V) alloy using abrasive water jet machining process. Procedia CIRP **46**, 139–142 (2016)
9. H. Li, J. Wang, An experimental study of abrasive waterjet machining of Ti-6Al-4V. Int. J. Adv. Manuf. Technol. **81**(1–4), 361–369 (2015)
10. V. Bhandarkar, R.A. Jibhakate, T.V.K. Gupta, Influence of process parameters on abrasive water jet machined pockets on Inconel 718 alloy, in *Smart Technologies for Energy, Environment and Sustainable Development*, ed. by M. Kolhe, P. Labhasetwar, H. Suryawanshi. Lecture Notes on Multidisciplinary Industrial Engineering (Springer, Singapore, 2019)
11. A.C. Arun Raj, S. Senkathir, T. Geethapriyan, J. Abhijit, Experimental investigation of abrasive waterjet machining of Nickel based superalloys (Inconel 625). IOP Conf. Ser. Mater. Sci. Eng. **402**(1), 012181 (2018)
12. M. Uthayakumar, M.A. Khan, S.T. Kumaran, A. Slota, J. Zajac, Machinability of nickel-based superalloy by abrasive water jet machining. Mater. Manuf. Processes **31**(13), 1733–1739 (2016)
13. T.U. Siddiqui, M. Shukla, P.B. Tambe, Optimisation of surface finish in abrasive water jet cutting of Kevlar composites using hybrid Taguchi and response surface method. Int. J. Mech. Mater. Des. **3**, 382–402 (2008)

14. S. Bhowmik, Jagadish, K. Gupta, *Modelling and Optimization of Advanced Manufacturing Processes* (Springer, 2019)
15. R. Ruiz-Garcia, P.F. Mayuet Ares, J.M. Vazquez-Martinez, J. Salguero Gómez, Influence of abrasive waterjet parameters on the cutting and drilling of CFRP/UNS A97075 and UNS A97075/CFRP Stacks. Materials **12**, 107 (2019)
16. K. Jayakumar, Abrasive water jet machining studies on Kenaf/E-glass fiber polymer composite, in *Proceedings of 10th International Conference on Precision, Meso, Micro and Nano Engineering*, Chennai, India, 7–9 Dec 2017, pp. 396–399
17. S. Madhu, M. Balasubramanian, Influence of nozzle design and process parameters on surface roughness of CFRP machined by abrasive jet. Mater. Manuf. Processes **32**(9), 1011–1018 (2017)
18. S. Xu, J. Wang, A study of abrasive waterjet cutting of alumina ceramics with controlled nozzle oscillation. Int. J. Adv. Manuf. Technol. **27**, 696 (2006)
19. D. Ghosh, P.K. Das, B. Doloi, Parametric studies of abrasive water jet cutting on surface roughness of silicon nitride materials, in *Proceedings of 5th International and 26th All India Manufacturing Technology, Design and Research Conference (AIMTDR 2014)*, Guwahati, India, 12–14 Dec 2014, pp. 422, 1–5
20. P.D. Unde, M.D. Gayakwad, N.G. Patil, R.S. Pawade, D.G. Thakur, P.K. Brahmankar, Experimental investigations into abrasive waterjet machining of carbon fiber reinforced plastic. J. Compos. ID 971596 (2015). https://doi.org/10.1155/2015/971596
21. M. Hashish, Abrasive waterjet cutting of microelectronic components, in *Proceedings of WJTA American Waterjet Conference* (2005)
22. V.K. Pal, S. Choudhury, Fabrication and analysis of micro-pillars by abrasive water jet machining. Procedia Mater. Sci. 61–71 (2014)
23. Z.W. Liu, R.Y. Liu, Study on pre-mixed micro abrasive water jet machining system. Appl. Mech. Mater. **618**, 475–479 (2014)
24. H. Orbanic, B. Jurisevic, D. Kramar, M. Grah, M. Junkar, Miniaturization of injection abrasive water jet machining process. Proc. IMechE Part C J. Mech. Eng. Sci. **220**, 1697–1705 (2006)
25. K.L. Pang, A study of the abrasive waterjet micro-machining process for amorphous glasses, PhD thesis, The University of New South Wales (2011)
26. T. Gutowski, J. Dahmus, S. Dalquist, Measuring the environmental load of manufacturing processes, in *International Society for Industrial Ecology (ISIE), 3rd International Conference on Industrial Ecology for a Sustainable Future*, Stockholm, Sweden (2003)
27. Jagadish, S. Bhowmik, A. Ray, Prediction of surface roughness quality of green abrasive water jet machining: a soft computing approach. J. Intell. Manuf. 1–15 (2015)
28. R. Kovacevic, M. Hashish, R. Mohan, M. Ramulu, T.J. Kim, E.S. Geskin, State of the art of research and development in abrasive water jet machining. J. Manuf. Sci. Eng. **119**(4B), 776–785 (1997)
29. T. Phokane, K. Gupta, M.K. Gupta, Near net shape manufacturing of miniature spur brass gears by abrasive water jet machining, in *Near Net Shape Manufacturing Processes*, ed. by K. Gupta (Springer, 2019), pp. 143–158
30. T.C. Phokane, K. Gupta, C. Popa, On abrasive water jet machining of miniature brass gears, in *Proceedings of International Gear Conference*, vol. II, Lyon (France) (Chartridge Books Oxford, 2018), pp. 384–392. ISBN 978-1-911033-43-1

Chapter 2
Abrasive Water Jet Machining of Metallic Materials

2.1 Introduction

Machining materials to the desired part geometry utilizing conventional machining (CV) processes (turning, milling, grinding, shaping, and so on) possess few limitations such as heat-affected zone and require designing cutting tool for every work material and applications [1]. In addition, complex part geometries coupled with high surface finish and dimensional accuracy may not be machined economically with CV processes [1]. The cutting fluids reduce the negative impact of heat generated at the cutting zone, tool life, residual stresses, and surface finish of CV processes [2]. The extensive benefits with cutting fluids must be balanced with the associated cost, health, and its disposal. Note that, cutting fluids account for approximately 7–17% of total machining cost [3]. The estimated cost may even rise to 30% for compensating the losses (vapor losses, leakage, loss with machine components, maintenance for cleaning and drying system) incurred while performing machining [4]. Furthermore, it is estimated that 80% of machining industries are facing serious health diseases due to the negative effect of cutting fluids [5]. Sustainable (environmental, social, and economically viable) machining methods are in great demand to limit the disadvantages of conventional machining methods [6].

In the last two decades, there has been a tremendous increase in research developments and innovations in the area of non-traditional or advanced machining techniques to find a suitable alternate of conventional machining processes. Advanced machining processes such as electrical discharge machining (EDM), laser beam machining (LBM), electron beam machining (EBM), and abrasive water jet machining (AWJM) offer significant benefits when machining complex geometries [1]. But on the other hand, there are certain inherent limitations of these processes. EDM possess a few challenges such as workpiece, and tool material must be a good conductor of electricity. In addition, EDM uses dielectric fluid to flush away the debris or chips formed from the cutting zone which not only accounts to additional cost but also generates the toxic fumes and gasses during machining [7]. LBM and EBM

Jagadish and K. Gupta, *Abrasive Water Jet Machining of Engineering Materials*,
Manufacturing and Surface Engineering,
https://doi.org/10.1007/978-3-030-36001-6_2

machining processes use thermal energy to cut parts possessing a wide range of industrial applications. Thermal energy machining processes undergo localized heating which causes thermal damage on the work material, which not only affects the assembly tolerances but also reduces the long-run performances as a result of poor surface finish [8, 9]. In view of the above, AWJM that can also be recognized green machining technology offers both industrial and societal benefits for the production of parts that are economically feasible by minimizing the pollutants and risk to human health and environment.

Green machining or sustainable machining technology offers significant benefits such as to conserve and improve natural ecosystem, reduce natural resources depletion, improve the waste management system, inhibit the generation of pollutants, and protect human health [10]. Green machining refers to processing raw materials to finish part geometry with minimum or no negative impact on the environment, personal health, productivity, and society [10, 11]. Green machining is also treated as an integral part of Industry 4.0 [12]. The green color generally represents the nature which in turn to plants and vegetables, whereas in manufacturing sectors green term refers to the environment or eco-friendly [13]. The AWJM performs machining under the influence of water and abrasive particles, which facilitates environmentally safe and eco-friendly machining process [14, 15]. The AWJM process finds major applications in machining almost all ductile and brittle materials (metal, polymers, plastics, wood, glass, stones, composites, ferrous and nonferrous metals) including difficult-to-cut materials (hardness above >45 HRC), thin sheets, foils, textiles, honeycomb, and leather materials usually by erosion [14]. Compared to other non-conventional processes, the AWJM technology offers significant advantages, such as minimized setup times; repeatedly, same tool can be used to perform machining different materials, comparatively faster machining, dustless machining which does not pollute environment and affects health, does not undergo plastic deformation and thermal stresses during machining and ability to cut thick-sectioned materials [15–18]. AWJM process is capable to cut complex geometries, but the process efficiency in terms of quality of machined surface, rate of material removal, energy consumption, and cost vary with respect to influencing process variables [19].

Hence, the present chapter put forward the experimental investigation of green abrasive water jet machining on AISI 304 grade steel material. For this, five process parameters, namely abrasive grain size (A), abrasive flow rate (B), nozzle speed (C), working pressure (D), and standoff distance (E), are used to know the green machining attributes like MRR, process time (PT), surface roughness (SR), and process energy (PE). Experimentation is done using Taguchi (L_{27}) orthogonal array to study the influence of each process parameters on the green machining parameters. Further, the effect of each machining parameters of AWJM on the green attributes or responses is analyzed using parametric analysis. In addition, regression analysis, ANOVA, and empirical models are done to show the statistical significance of the green machining process and for optimum prediction of the green attribute of AWJM process, respectively. Finally, confirmatory tests are performed to verify results with experimental results.

2.2 Material and Method

2.2.1 Material Specimen

AISI 304 steel material possessing 10-mm-thick plate was used for the present work. The chemical composition of AISI 304 steel material is listed in Table 2.1. The AISI 304 steel material mechanical and physical properties are presented in Table 2.2. Grade 304 is the most versatile and widely used stainless steel possessing excellent forming and welding characteristics. In addition, excellent corrosion resistance showed the materials used for a wide range of products and forms. The stainless steels (AISI 304) also possess a greater ability to withstand strength at elevated temperature and offers greater resistance to oxidation. Typically, AISI 304 showed numerous applications in aerospace, automotive (heat exchangers, threaded fasteners, springs, engine parts, etc.), electronic industries, and food processing equipment's (such as beer brewing, winemaking, and milk processing).

Table 2.1 Chemical composition of AISI 304 steel material

Material composition	Value
C	0.08
Mn	2
Si	1
Cr	18–20
Ni	0.8–12
P	0.045
Si	0.03

Table 2.2 Mechanical and physical properties of AISI 304 steel material

Mechanical properties	Units	Value	Physical properties	Units	Value
Tensile strength	MPa	515	Density	kg/m^3	8000
Yield strength of 0.2%	MPa	205	Poisson's ratio	–	0.27–0.3
Elongation	%	40	Elastic modulus	GPa	193
Rockwell hardness	HRB	96	Thermal conductivity at 100 °C	W/m K	16.2
Brinell hardness	HB	150	Specific heat 0–100 °C	J/kg K	150
			Electrical resistivity	nΩ m	720

2.2.2 Experimental Details

The CNC Abrasive Water Jet Cutting Machine (make: DARDI International Corporation Ltd., China) also know a green machining process was used to conduct experiments and collect input–output data (refer Fig. 2.1).

The machine-related parameters such as designed pressure, discharge rate, and orifice diameter were maintained to a constant value of 3800 bar, 2.31 l/min, and 0.25 mm, respectively. Abrasive slurry was prepared under room atmosphere conditions, by mixing with an appropriate combination of distilled water and the Garnet abrasive material (80 [$\approx$240 μm]). During experimentation, the inputs of the AWJM machine such as voltage, current, and nozzle angle are kept fixed to 300 V, 20 A, and 90°, respectively. Experiments were conducted on the work specimen of size possessing the dimension of 300 mm × 115 mm × 10 mm. In the present work, five parameters such as abrasive material grain size, standoff distance, working pressure, nozzle speed, and abrasive mass flow rate have been varied at three levels (Table 2.3), and experiments have been designed using Taguchi (L27) orthogonal array. A square

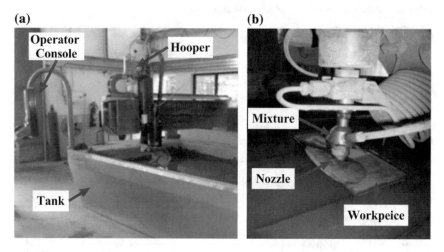

Fig. 2.1 **a** AWJM experimental setup, **b** AWJM nozzle head system

Table 2.3 Machining parameters and their levels of green machining (i.e., AWJM) process

Control factors	Units	Levels (1, 2 and 3)	Notation
Abrasive grain size	mesh	60, 80 and 100	A
Standoff distance	mm	1.5, 2.5 and 3.5	B
Working pressure	MPa	150, 225 and 300	C
Nozzle speed	mm/min	125, 175 and 225	D
Abrasive mass flow rate	g/s	3, 5 and 7	E

through cut of 20 by 20 mm^2 was made during each experiment. Each experiment was performed three times and their average values of manufacturing parameters like MRR, SR, and process time (PT) and environmental parameters like process energy (PE) in the analysis (Table 2.4).

During experimentation, response or green parameters (i.e., MRR, SR, PT, and PE) were evaluated using the following expressions:

$$MRR = \frac{W_i - W_f}{C_T} \tag{2.1}$$

$$PT = \frac{60}{MRR} \tag{2.2}$$

$$PE = PT \times V \times I \tag{2.3}$$

where W_i and W_f are initial weight and the final weight; C_T is the machining time in min; V is the voltage in volt, and I represent current supplied in amp; MRR is the material removal rate, and PE is the process energy.

Thereafter, machined surfaces were measured using surface profilometer (make: Tokyo Seimitsu Co. Ltd. Model: Handysurf E-35B) to determine the SR. At last, the collected output parameters were transformed into S/N ratio data depending on the desired quality characteristics. Note that, the larger-the-better criterion is selected for the green parameter, i.e., MRR and smaller-the-better green parameters, i.e., PT, PE, and SR, respectively. Table 2.5 presents the S/N ratio values of the response or green parameters correspond to AWJM process.

$$S/N_{\text{Larger-the-better}} = -10 \log \frac{1}{n}(y^2) \tag{2.4}$$

$$S/N_{\text{Smaller-the-better}} = -10 \log \frac{1}{n}({1}/{y^2}) \tag{2.5}$$

2.3   Results and Discussion

2.3.1 Parametric Analysis

Figure 2.2 depicts the influence of variation in main machining parameters, i.e., A, B, C, D, and E on green responses like MRR, PT, SR, and PE of AWJM process was studied. It has been observed that an increase in mesh number or decreasing the size (i.e., diameter) of the abrasive particles the surface quality improves drastically. This is attributed to the coarse size (mesh 60) abrasive particle resulted in faster cut and generate rough surface [20]. Coarse size (large size) abrasive particles hit the work material at greater impact force which tends to remove the surface of the material

Table 2.4 Experimental results of green machining (i.e., AWJM) process

Exp. No.	Machining parameters					Response (green) parameters			
	A (mesh)	B (mm)	C (MPa)	D (mm/min)	E (g/s)	SR (μm)	MRR (mm³/min)	PT (s)	PE (W)
1	60	1.5	150	125	3	2.00	534	0.112	674.2
2	60	1.5	150	125	5	1.97	490	0.122	734.7
3	60	1.5	150	125	7	1.91	423	0.142	851.1
4	60	2.5	225	175	3	1.97	432	0.139	833.3
5	60	2.5	225	175	5	1.93	412	0.146	873.8
6	60	2.5	225	175	7	1.92	321	0.187	1121.5
7	60	3.5	300	225	3	1.90	381	0.157	944.9
8	60	3.5	300	225	5	1.92	431	0.139	835.3
9	60	3.5	300	225	7	1.91	426	0.141	845.1
10	80	1.5	225	225	3	1.55	286	0.210	1258.7
11	80	1.5	225	225	5	1.66	387	0.155	930.2
12	80	1.5	225	225	7	1.42	342	0.175	1052.6
13	80	2.5	300	125	3	1.14	198	0.303	1818.2
14	80	2.5	300	125	5	1.33	243	0.247	1481.5
15	80	2.5	300	125	7	1.03	199	0.302	1809.0
16	80	3.5	150	175	3	1.77	392	0.153	918.4
17	80	3.5	150	175	5	1.73	385	0.156	935.1
18	80	3.5	150	175	7	1.65	296	0.203	1216.2
19	100	1.5	300	175	3	1.30	268	0.224	1343.3
20	100	1.5	300	175	5	1.35	306	0.196	1176.5

(continued)

Table 2.4 (continued)

Exp. No.	Machining parameters					Response (green) parameters			
	A (mesh)	B (mm)	C (MPa)	D (mm/min)	E (g/s)	SR (µm)	MRR (mm³/min)	PT (s)	PE (W)
21	100	1.5	300	175	7	1.08	199	0.302	1809.0
22	100	2.5	150	225	3	1.54	341	0.176	1055.7
23	100	2.5	150	225	5	1.49	284	0.211	1267.6
24	100	2.5	150	225	7	1.09	232	0.259	1551.7
25	100	3.5	225	125	3	1.35	304	0.197	1184.2
26	100	3.5	225	125	5	1.24	256	0.234	1406.3
27	100	3.5	225	125	7	1.05	208	0.288	1730.8

Table 2.5 S/N ratio values of green machining (AWJM) process

Exp. No.	SR (μm)	MRR (mm³/min)	PT (s)	PE (W)	Exp. No.	SR (μm)	MRR (mm³/min)	PT (s)	PE (W)
1	−6.003	54.55	18.99	−56.58	15	−0.290	45.98	10.41	−65.15
2	−5.903	53.80	18.24	−57.32	16	−4.935	51.87	16.30	−59.26
3	−5.612	52.53	16.96	−58.60	17	−4.771	51.71	16.15	−59.42
4	−5.889	52.71	17.15	−58.42	18	−4.355	49.43	13.86	−61.70
5	−5.707	52.30	16.73	−58.83	19	−2.265	48.56	13.00	−62.56
6	−5.671	50.13	14.57	−61.00	20	−2.632	49.71	14.15	−61.41
7	−5.561	51.62	16.06	−59.51	21	−0.677	45.98	10.41	−65.15
8	−5.684	52.69	17.13	−58.44	22	−3.739	50.66	15.09	−60.47
9	−5.630	52.59	17.03	−58.54	23	−3.470	49.07	13.50	−62.06
10	−3.784	49.13	13.56	−62.00	24	−0.725	47.31	11.75	−63.82
11	−4.386	51.75	16.19	−59.37	25	−2.626	49.66	14.09	−61.47
12	−3.015	50.68	15.12	−60.45	26	−1.875	48.16	12.60	−62.96
13	−1.146	45.93	10.37	−65.19	27	−0.449	46.36	10.80	−64.76
14	−2.464	47.71	12.15	−63.41					

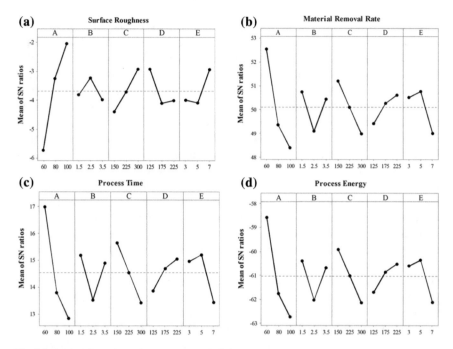

Fig. 2.2 Main effect plots of responses: **a** SR, **b** MRR, **c** PT, and **d** PE

by generating more damaged surface texture. It was also observed that the surface finish improves up to 2.5 mm of standoff distance and thereafter showed decreasing trends. Important to note that, the influence of standoff distance on surface roughness is found negligible (refer Fig. 2.2). The kinetic energy density of the jet at the point of impact with the workpiece material reduces as a result of increased jet diameter with the presence of air drag [21]. An increase in working pressure causes increased abrasive fragmentation within the nozzle and prevents spreading water leaving the jet, which favors toward positive impact on surface quality. The similar trend results were observed in the research investigations by Kovacevic [20]. Increased abrasive mass flow rate tends to decrease the surface roughness might be due to the increased interface between a large number of abrasive particles reduces the impact force as a result of altered impact angle and reduced velocities [22]. There will be an observed increasing trend of surface roughness with an increase in nozzle speed (refer Fig. 2.2). This occurs due to the kinetic energy of the abrasive particle that improves with the negligible overlap, which in turn possesses higher impact force and thereby reduces surface quality [14].

The main effect plots of input variables on the responses (MRR, PT, and PE) are presented in Fig. 2.2. It was observed that low values of grain size of abrasive particle resulted in higher MRR, low process time, and process energy. This is attributed to the smaller mesh (i.e., 60) size resulted in coarse-sized abrasive particles which possess major impact that covers or hit the larger surface area on which the material to cut causes higher MRR. The trend of abrasive grain size over material removal rate is found to be in good agreement with the results of machining aluminum alloy [23]. It was observed that nozzle speed and standoff distance have less impact on the material removal rate. Note that, with increase in standoff distance the material removal rate decreases due to spreading of a water jet in the air as the distance accelerates tend to break up the water in the form of droplets results in decreased erosion rate. MRR showed an increasing trend with traverse rate due to the increased intermolecular forces and energy causes to shear initially the work material leads to erosion from the surface to be machined [24]. As the working pressure increases, the abrasive particle fragmentation increases within the nozzle as a result of increased water flow rate. The breaking of abrasive particles to smaller size results in low impact forces and contact surfaces. Increase in abrasive flow rate tends to increase the kinetic energy of the water jet results in deeper penetration of the abrasive particles tends to form initial cracks followed by removal of work material in the form of chips resulted in higher MRR [25, 26]. It was also observed that beyond the critical abrasive flow rate, the material removal tends to decrease, which might be due to the fragmentation of abrasive particles at the higher kinetic energy of water flow rate.

The PT and PE were computed based on the material removal rate data information. Therefore, the main effect plots of various input factors obtained for MRR remain identical with the PT and PE (refer Fig. 2.2a–d). However, to optimize process according to industry requirements the material removal rate must be maximized while minimizing the processing time and energy consumption. Therefore, MRR has shown an inverse relationship with PT and PE. Thereby, conflicting requirements need to be met while optimizing multiple objective functions.

2.3.2 Regression Analysis

Additionally, this research also carried out the regression analysis and ANOVA for statistical significance. The effect of machining parameters on the response (green) parameters MRR, SR, PT, and PE is determined by developing an empirical model for predicting the response values of AWJM process. The regression equations for response (green) parameters like MRR, SR, PT, and PE are listed in Eqs. (2.6)–(2.9), respectively.

$$MRR = 816.3 - 4.033A - 8.7B - 0.538C + 0.283D - 13.61E \qquad (2.6)$$

$$PT = -0.0486 + 0.002227A + 0.00172B + 0.000353C \\ - 0.000361D + 0.00906E \qquad (2.7)$$

$$PE = -292 + 13.36A + 10.3B + 2.117C - 2.16D + 54.3E \qquad (2.8)$$

$$SR = 3.122 - 0.01648A + 0.0167B - 0.001609C + 0.01600D - 0.0401E$$

$$(2.9)$$

wherein Eq. (2.6) for MRR shows that parameter nozzle speed (D) poses a positive effect for MRR while other parameters pose negative effects. In Eq. (2.7), for PT shows that abrasive material grain size (A), stand of distance (B), working pressure (C), and abrasive mass flow rate (E) poses positive effect, and parameter nozzle speed (D) poses a negative effect on PT. Similarly, the same pattern of effect is found for PE, i.e., abrasive material grain size (A), stand of distance (B), working pressure (C), and abrasive mass flow rate (E) poses positive effect, and parameter nozzle speed (D) poses a negative effect on PE. On the other hand, for SR (Eq. 2.9) shows that the abrasive material grain size, working pressure, and abrasive mass flow rate have a negative influence, whereas standoff distance and nozzle speed poses positive effect with surface roughness.

The factor that influences toward statistical significance and model adequacies developed were tested for the preset confidence level set at 95% utilizing analysis of variance (ANOVA). In ANOVA, the coefficient of determination (R^2) and adjusted R^2 is determined to know the developed model statistical adequacies. Fisher's statistical test (F ratio) and statistical probability (P-value) are used to know the most influencing parameters and the statistical significance toward response variables, respectively. Larger F-statistic value of a parameter dictates the most significant parameter, while the P value ($P < 0.5$) determines the parameter that is statistically significant. The R^2 value signifies the ratio of the sum of squares of regression and the total sum of squares which explain the variation in the response. The values of R^2 vary in the ranges from 0 to 1. Larger the R^2 value (i.e., close to 1) corresponds to a more statistically significant model. Excluding insignificant terms by the model through backward elimination method to fit the regression model signifies the adjusted R^2 value. Note that, the non-contributing terms need not be removed from the model, as it not only resulted in an imprecise input–output relationship but also reduces the prediction accuracy.

The parameter significance and percent contribution of factors tested on the different responses (MRR, PT, and PE) by utilizing the analysis of variance and the results are provided in Tables 2.6 and 2.7, respectively. The result shows that abrasive material grain size (A) found to have major contribution followed by working pressure (C), abrasive mass flow rate (E), standoff distance (B), and nozzle speed (D). Note that, the influence of nozzle speed (D) is found to have a negligible impact (as their P-value is less than 0.05) on MRR, PT, and PE. The coefficient of determination and adjusted R^2 value was found equal to 0.891 and 0.823, respectively. Therefore, the models developed for the responses (PE, PT, and MRR) are found to be statistically adequate for prediction and optimization. The optimal levels for the responses (MRR, PE, and PT) are found to be $A_1B_1C_1D_3E_2$.

Similarly, the percent contribution of parameters (A, B, C, D, and E) toward SR is estimated. The parameter abrasive material grain size (A) resulted in the highest contribution as their F-value and sequential sum of squares are found to be higher. It is important to note that, the abrasive material grain size followed by working pressure

Table 2.6 ANOVA for different responses—SR, MRR, PT, and PE

Details	DF	Seq. SS	Adj. MS	F-value	P-value	% contribution
Responses	*SR*					
Abrasive material grain size	2	63.83	31.9148	80.31	0.000	65.37
Standoff distance	2	2.794	1.3971	3.52	0.054	2.86
Working pressure	2	9.640	4.8202	12.13	0.001	9.87
Nozzle speed	2	7.573	3.7865	9.58	0.002	7.76
Abrasive mass flow rate	2	7.455	3.7274	9.38	0.002	7.63
Residual error	16	6.358	0.3974			6.51
Total	26	97.65				
Responses	*MRR, PE, PT*					
Abrasive material grain size	2	85.320	42.660	38.47	0.000	52.31
Standoff distance	2	14.077	7.038	6.35	0.009	8.63
Working pressure	2	22.536	11.268	10.16	0.001	13.82
Nozzle speed	2	6.876	3.438	3.10	0.073	4.22
Abrasive mass flow rate	2	16.546	8.273	7.46	0.005	10.14
Residual error	16	17.743	1.109			10.88
Total	26	163.097				

(C), nozzle speed (D), and abrasive mass flow rate (E) is statistically significant listed according to their paramount importance (refer Table 2.6). Note that, the P-value of standoff distance is found greater than 0.05 (i.e., insignificant), which indicates their influence on surface roughness is negligible (refer Table 2.6). Furthermore, the R^2 values (including all parameters) that correspond to surface roughness were found equal to 0.891 and the adjusted R^2 (excluding standoff distance) value was equal to 0.823. The above tests proved to be an effective tool for the developed model toward this response. The optimal levels for the surface roughness are found to be $A_3B_2C_3D_1E_3$.

2.3.3 Modeling and Optimization

The modeling of AWJM under manufacturing and environmental scenarios is done using DEAR method [14]. Work optimized the AWJM parameters under two conditions: first—determination of optimal conditions considering manufacturing parameters like MRR, SR, and PT and second—determination optimal conditions considering environmental parameters like PE. In optimization, response/output

Table 2.7 Results of ANOVA of different responses

Level	A	B	C	D	E	A	B	C	D	E
Response	SR					MRR				
1	−5.740	−3.815	−4.396	−2.922	−4.003	52.55	50.74	51.21	49.41	50.52
2	−3.244	−3.233	−3.713	−4.099	−4.094	49.35	49.09	50.10	50.27	50.77
3	−2.046	−3.981	−2.921	−4.009	−2.933	48.39	50.45	48.97	50.61	49.00
Delta	3.694	0.748	1.475	1.177	1.161	4.16	1.66	2.24	1.20	1.77
Optimal levels	$A_3B_2C_3D_1E_3$					$A_1B_1C_1D_3E_2$				
Response	PT					PE				
1	16.98	15.18	15.65	13.85	14.96	−58.58	−60.38	−59.91	−61.72	−60.61
2	13.79	13.52	14.54	14.70	15.21	−61.77	−62.04	−61.03	−60.86	−60.36
3	12.82	14.89	13.41	15.05	13.43	−62.74	−60.67	−62.15	−60.52	−62.13
Delta	4.16	1.66	2.24	1.20	1.77	4.16	1.66	2.24	1.20	1.77
Optimal levels	$A_1C_1D_3E_2$					$A_1B_1D_3E_2$				

parameters like MRR, SR, PT, and PE are considered as criteria while experimental runs which contain the input parameter [abrasive material grain size (A), standoff distance (B), the working pressure (C), nozzle speed (D), and abrasive mass flow rate (E)] variation alternatives.

During optimization, first, the development of decision matrix is done which includes a number of criteria as response parameters and a number of operating/input parameter setting as alternatives. The experimental result as provided in Table 2.4 is taken as a decision matrix in this case. Second, the determination of weights for each of the AWJM process parameters is done using Eqs. (2.10–2.13) [27, 28], and the results are provided in Table 2.8.

$$W_{MRR} = \frac{MRR}{\sum MRR} \tag{2.10}$$

$$W_{SR} = \frac{\left(^1/_{SR}\right)}{\sum \left(^1/_{SR}\right)} \tag{2.11}$$

$$W_{PT} = \frac{\left(^1/_{PT}\right)}{\sum \left(^1/_{PT}\right)} \tag{2.12}$$

$$W_{PE} = \frac{\left(^1/_{PE}\right)}{\sum \left(^1/_{PE}\right)} \tag{2.13}$$

After that, the formulation of weighted decision matrix and determination of MPRI values for each of the AWJM output parameter is done using Eqs. (2.6–2.8) [27, 28], and the results are provided in Table 2.8.

$$MPRI = \frac{M}{(S + P + T)} \tag{2.14}$$

Thereafter, percent (%) contribution of individual parameters is determined with abrasive grain size (66.51%) being the most dominated factor followed by working pressure (14.46%), abrasive mass flow rate (8.83%), standoff distance (9.90%), and nozzle speed (0.31%), respectively. This indicates the nozzle speed found to have least significant factor considering all output functions. Note that, $A_1B_1C_1D_3E_2$ be the optimal levels for the abrasive water jet machining process to improve the performances of multiple outputs (refer Table 2.9).

Table 2.8 DEAR optimization results of AWJM process

Exp. No.	Response (green) parameters				Weighted normalized value				MPRI
	1/SR	MRR	1/PT	1/PE	WSR	WMRR	WPT	WPE	
1	0.500	534	8.929	0.0015	0.028	0.060	0.061	0.060	0.791
2	0.508	490	8.197	0.0014	0.028	0.055	0.056	0.055	0.584
3	0.524	423	7.042	0.0012	0.029	0.048	0.048	0.048	0.496
4	0.508	432	7.194	0.0012	0.028	0.049	0.049	0.049	0.618
5	0.518	412	6.849	0.0011	0.029	0.047	0.046	0.047	0.431
6	0.521	321	5.348	0.0009	0.029	0.036	0.036	0.036	0.286
7	0.526	381	6.369	0.0011	0.029	0.043	0.043	0.043	0.352
8	0.521	431	7.194	0.0012	0.029	0.049	0.049	0.049	0.515
9	0.524	426	7.092	0.0012	0.029	0.048	0.048	0.048	0.503
10	0.645	286	4.762	0.0008	0.036	0.032	0.032	0.032	0.190
11	0.602	387	6.452	0.0011	0.033	0.044	0.044	0.044	0.415
12	0.704	342	5.714	0.0010	0.039	0.039	0.039	0.039	0.370
13	0.877	198	3.300	0.0005	0.048	0.022	0.022	0.022	0.042
14	0.752	243	4.049	0.0007	0.041	0.027	0.027	0.027	0.182
15	0.971	199	3.311	0.0006	0.054	0.023	0.022	0.023	0.110
16	0.565	392	6.536	0.0011	0.031	0.044	0.044	0.044	0.433
17	0.578	385	6.410	0.0011	0.032	0.044	0.044	0.044	0.411
18	0.606	296	4.926	0.0008	0.033	0.033	0.033	0.033	0.243
19	0.769	268	4.464	0.0007	0.042	0.030	0.030	0.030	0.166
20	0.741	306	5.102	0.0008	0.041	0.035	0.035	0.035	0.297

(continued)

Table 2.8 (continued)

Exp. No.	Response (green) parameters					Weighted normalized value				
	1/SR	MRR	1/PT	1/PE	WSR	WMRR	WPT	WPE	MPRI
21	0.926	199	3.311	0.0006	0.051	0.023	0.022	0.023	0.099
22	0.649	341	5.682	0.0009	0.036	0.039	0.039	0.039	0.367
23	0.671	284	4.739	0.0008	0.037	0.032	0.032	0.032	0.224
24	0.917	232	3.861	0.0006	0.051	0.026	0.026	0.026	0.076
25	0.741	304	5.076	0.0008	0.041	0.034	0.034	0.034	0.344
26	0.806	256	4.274	0.0007	0.045	0.029	0.029	0.029	0.148
27	0.952	208	3.472	0.0006	0.053	0.024	0.024	0.024	0.107

Table 2.9 Pareto ANOVA for the combined responses (SR, MRR, PE, and PT): DEAR

Factors	Levels	A	B	C	D	E	Total
The sum at factor levels	1	**4.6484**	**3.4885**	**3.7289**	2.8935	3.2509	8.900
	2	2.4293	2.3520	2.7989	2.9232	**3.3074**	
	3	1.8223	3.0595	2.3722	**3.0833**	2.3417	
Sum of squares of differences		13.280	1.976	2.887	0.063	1.762	19.968
Percent contribution		66.51	9.90	14.46	0.31	8.83	100
Optimal levels		$A_1B_1C_1D_3E_2$					

Bold values indicates: *Optimal Values

Table 2.10 Confirmatory results

Models	Optimal levels	Optimal machining parameters	Optimal response (green) parameters
DEAR	$A_1B_1C_1D_3E_2$	A: 60 mesh	SR: 1.84 μm
		B: 1.5 mm	MRR: 468 mm^3/min
		C: 150 MPa	PT: 0.128 s
		D: 225 mm/min	PE: 769 W
		E: 5 g/s	

2.3.4 Confirmation Experiments

After the optimization, work is also carried out the confirmatory analysis/experimentation to confirm the results obtained via DEAR method. The confirmatory tests are performed based on the optimal setting obtained using DEAR method, and the confirmatory results are provided in Table 2.10. Important to note that the optimal levels recommended by the DEAR method are not among the combination of L_{27} orthogonal array experiments of Table 2.4. This occurs due to the multifactor nature of Taguchi experimental design (i.e., $3^5 = 243$). It is also observed that optimal values obtained via confirmatory experiments found acceptable and satisfactory with that of the experimental results.

2.4 Summary

The sustainable machining method (i.e., AWJM process) for machining of AISI 304 steel is presented in this research work. Taguchi L_{27} orthogonal array experiments are conducted for to study the most influencing variables [abrasive grain size (A), nozzle speed (B), abrasive mass flow rate (C), working pressure (D) and standoff distance (E)] on the green parameters (MRR, SR, PE and PT). Based on the experimental results, parametric analysis, regression analysis, and optimization following conclusions are drawn from the present work.

- From the parametric analysis, abrasive grain size (A) showed the most dominating effect compared to the other parameters for ANSI 300 stainless steel. In order to get optimal performance parameter, abrasive grain size needs to be set lower level during the machining of C in AWJM process.
- Optimal setting $A_3B_2C_3D_1E_3$ for SR and $A_1B_1C_1D_3E_2$ for MRR, PT, and PE are obtained for AWJM (green machining) process.
- The overall optimal setting obtained is $A_1B_1C_1D_3E_2$, i.e., A (60 mesh, level 1), B (1.5 mm, level 1), C (150 MPa, level 1), D (225 mm/min, level 3), and E (5 g/s, level 2). The corresponding green attributes obtained are SR as 1.84 μm, MRR as 468 mm^3/min, PT as 0.128 s and PE as 769 W.
- The optimal setting provides optimal responses such as the higher MRR, lesser PT, smoother SR, less consumption PE which has less influence on a generation of environmental issues aroused during the machining of ANSI 300 stainless steel in AWJM (green machining) process.
- Statistical significance of the data is done via ANOVA and regression analysis. It is found that the results are found to be statistically significant and follow a normal distribution and the normality assumption is valid.
- Confirmatory results for green parameters like MRR, SR, PT, and PE are found closer to the experimental results and well within the considerable ranges and satisfactory.

Finally, it is concluded that the AWJM a green machining process has potential for machining of AISI 300 stainless steel under green machining environment. The developed models and optimal parameter setting can be used for future references while machining ANSI 300 stainless steel.

References

1. G. Kibria, B. Bhattacharyya, J.P. Davim, *Non-traditional Micromachining Processes* (Springer, 2017)
2. E. Kuram, B. Ozcelik, E. Demirbas, Environmentally friendly machining: vegetable based cutting fluids, in *Green Manufacturing Processes and Systems* (Springer, Berlin, Heidelberg, 2013), pp. 23–47
3. D.P. Adler, W.W.S. Hii, D.J. Michalek, J.W. Sutherland, Examining the role of cutting fluids in machining and efforts to address associated environmental/health concerns. Mach. Sci. Technol. **10**(1), 23–58 (2006)
4. G. Byrne, E. Scholta, Environmentally clean machining processes—a strategic approach. CIRP Ann. Manuf. Technol. **42**(1), 471–474 (1993)
5. Y.M. Shashidhara, S.R. Jayaram, Vegetable oils as a potential cutting fluid—an evolution. Tribol. Int. **43**(5–6), 1073–1081 (2010)
6. J.P. Davim, Non-traditional machining processes, in *Manufacturing Process Selection Handbook* (2013), pp. 205–226

7. Y.C. Lin, J.C. Hung, H.M. Chow, A.C. Wang, J.T. Chen, Machining characteristics of a hybrid process of EDM in gas combined with ultrasonic vibration and AJM. Procedia CIRP **42**, 167–172 (2016)
8. C.T. Pan, H. Hocheng, Laser machining and its associated effects, in *Advanced Analysis of Non-traditional Machining* (Springer, New York, NY, 2013), pp. 1–64
9. M. Ramulu, D. Arola, Water jet and abrasive water jet cutting of unidirectional graphite/epoxy composite. Composites **24**(4), 299–308 (1993)
10. D. Fratila, Sustainable manufacturing through environmentally-friendly machining, in *Green Manufacturing Processes and Systems* (Springer, Berlin, Heidelberg, 2013), pp. 1–21
11. K. Gupta, R.F. Laubscher, J.P. Davim, N.K. Jain, Recent developments in sustainable manufacturing of gears: a review. J. Clean. Prod. **112**, 3320–3330 (2016)
12. Y. Meng, Y. Yang, H. Chung, P.H. Lee, C. Shao, Enhancing sustainability and energy efficiency in smart factories: a review. Sustainability **10**(12), 4779 (2018)
13. U.S. Dixit, D.K. Sarma, J.P. Davim, *Environmentally Friendly Machining* (Springer Science & Business Media, 2012)
14. J. Wang, W.C.K. Wong, A study of abrasive waterjet cutting of metallic coated sheet steels. Int. J. Mach. Tools Manuf. **39**(6), 855–870 (1999)
15. T.C. Phokane, K. Gupta, Sustainable manufacturing of precision miniature gears by abrasive water jet machining-an experimental study, in *Proceedings: 15th International Conference on Manufacturing Research (ICMR)* (2017)
16. A.W. Momber, R. Kovacevic, *Principles of Abrasive Water Jet Machining* (Springer Science & Business Media, 2012)
17. J.Y. Sheikh-Ahmad, *Machining of Polymer Composites* (Springer, New York, 2009), pp. 164–165
18. V. Gupta, P.M. Pandey, M.P. Garg, R. Khanna, N.K. Batra, Minimization of kerf taper angle and kerf width using Taguchi's method in abrasive water jet machining of marble. Procedia Mater. Sci. **6**, 140–149 (2014)
19. M. Shukla, Abrasive water jet milling, in *Non-traditional Machining Processes* (Springer, London, 2013), pp. 177–203
20. R.A.D.O.V.A.N. Kovacevic, Surface texture in abrasive waterjet cutting. J. Manuf. Syst. **10**(1), 32–40 (1991)
21. Jagadish, K. Gupta, M. Rajakumaran, Evaluation of machining performance of pineapple filler based reinforced polymer composites using abrasive water jet machining process. IOP Conf. Ser. Mater. Sci. Eng. **430**(1), 012046. IOP Publishing (2018)
22. M. Hashish, A modeling study of metal cutting with abrasive waterjets. J. Eng. Mater. Technol. **106**(1), 88–100 (1984)
23. A. Akkurt, The effect of cutting process on surface microstructure and hardness of pure and Al 6061 aluminium alloy. Eng. Sci. Technol. **18**(3), 303–308 (2015)
24. K.R. Kumar, V.S. Sreebalaji, T. Pridhar, Characterization and optimization of abrasive water jet machining parameters of aluminium/tungsten carbide composites. Measurement **117**, 57–66 (2018)
25. R.H.M. Jafar, H. Nouraei, M. Emamifar, M. Papini, J.K. Spelt, Erosion modeling in abrasive slurry jet micro-machining of brittle materials. J. Manuf. Process. **17**, 127–140 (2015)
26. M. Santhanakumar, R. Adalarasan, M. Rajmohan, Parameter design for cut surface characteristics in abrasive waterjet cutting of $Al/SiC/Al_2O_3$ composite using grey theory based RSM. J. Mech. Sci. Technol. **30**(1), 371–379 (2016)
27. F. Cavallaro, Multi-criteria decision aid to assess concentrated solar thermal technologies. Renew. Energy **34**(7), 1678–1685 (2009)
28. P. Aragonés-Beltrán, F. Chaparro-González, J.P. Pastor-Ferrando, A. Pla-Rubio, An AHP (Analytic Hierarchy Process)/ANP (Analytic Network Process)-based multi-criteria decision approach for the selection of solar-thermal power plant investment projects. Energy **66**, 222–238 (2014)

Chapter 3
Abrasive Water Jet Machining of Polymer Composites

3.1 Introduction

There has been an increased demand of polymer composites in various application segments such as precision engineering, scientific, aerospace, automotive, tool and die making, and household applications [1, 2]. To fulfill the demand, various forms of polymer composites have been fabricated such as glass fiber reinforced polymer (GFRP) composite, ceramic matrix composites (CMC), carbon fiber reinforced polymer (CFRP) composite, natural fiber polymer (NFRP) composite [3]. From these, NFRP composite is popular due to their greater properties of sustainability and eco-friendly nature. The fiber and filler used for NFRP composite obtained from various sources, i.e., pineapple, banana, flax, cotton, husk, bamboo, hemp, jute, and wood [4]. Instead of this wood dust or wood filler-based composite is very cheap and has larger industrial application. However, for large scale of application, machining of wood dust filler reinforcement polymer (WDFRP) or green composite (GC) is highly essential [5]. The machining in the conventional technique and some of the non-conventional machining create long-lasting problems. Due to their intermolecular structure, fibers pull out during drilling, poor surface roughness, improper cutting, low productivity, etc. In addition, conventional or some non-conventional processes generate various forms of wastages in the form of solid, liquid, and gaseous wastages which result in serious occupational health and environmental issues during the machining process. Hence, environmental friendly non-conventional machining processes are essential [6]. At present, various non-conventional or advanced machining processes are available, namely laser cutting or laser beam machining (LBM), ultrasonic machining (USM), wire-EDM (WEDM), and abrasive water jet machining, for machining of various engineering materials [7]. The machining of WDFRP or green composite in EDM and WEDM is not applicable [8] due to the non-conductive property of the material. Further, due to more ductility and low hardness of the WDFRP composites, machining in USM is also not possible [9]. In the case of LBM on WDFRP composites, the temperature of the beam badly affects mate-

Jagadish and K. Gupta, *Abrasive Water Jet Machining of Engineering Materials*,
Manufacturing and Surface Engineering,
https://doi.org/10.1007/978-3-030-36001-6_3

rial surface and this process burns the edges of cutting holes [10]. The AWJM is a preferable method for machining of WDFRP composite material because the process offers many benefits such as no thermal effect; impose low stress on the workpiece, higher versatility and higher flexibility, less environmental contamination, no fumes or aerosols generation, no wastage generation, hence AWJM sometime called as environment friendly or green machining or green manufacturing process [5, 11].

The various researchers worked on the polymer composite but very few researches done on machining of polymer composite specifically green composites under green machining environment, i.e., using AWJM process. Oksman and Selin [12] investigated the elastic modulus of wood fiber which is approximately 40 times than that of polyethylene and the strength about 20 times. Shaikh and Jain [13] found that the problem of fiber pulls out, damage of work specimen in case of diamond saw cutting while fiber curling and pulling in multiple directions with moisturizing effect of natural fiber in case of water jet machining. Lemma et al. [14] studied the effect of frequency of oscillation and the angle of oscillation on improving surface quality of glass fiber reinforcement polymer (GFRP) composite in AWJM process. Shanmugam et al. [15] examined the machinability of two composites, i.e., carbon composite and fiber reinforced plastic for investigating the kerf characteristics and surface roughness while machining by AWJM, plain water jet machining and laser cutting. Ke et al. [16] performed the machinability study of novel composites, i.e., silicon wafer in abrasive water jet machining. Their investigation result identified AWJM has potential to machine novel composites and obtains good surface quality. Sasikumar et al. [17] employed AWJM process for machining of hybrid aluminum 7075 metal matrix composites to study dimensional deviation, i.e., kerf. Three different percentage reinforcements, i.e., of 5, 10, and 15% of TiC and B_4C (equal amount of each), were employed and four independent parameters such as water jet pressure, jet traverses speed, and standoff distance were studied to get optimal results. Kalirasu et al. [18] investigated the mechanical and machining performance of glass and coconut sheath fiber–polyester composites using AWJM process. The work used Taguchi design for experimental design taking three independent parameters like a stand of distance, abrasive particle size, and cutting force on two dependent parameters such as kerf taper angle and surface roughness. It was observed that machining of coconut sheath fiber–polyester composites shows better machinability compare to the glass composites. However, there is a scarcity of work on AWJM of WDFRP composites.

Therefore, the present chapter aims to conduct experimental investigation on machining on green composites, i.e., WDFRP composites by abrasive water jet process. For this, four process parameters, namely standoff distance (SoD), working pressure (WP), nozzle speed (NS), and abrasive grain size (AGS), are used to know the green machining attributes like MRR, surface roughness (SR), and machining time (MT). Experimentations are done using Taguchi (L_{09}) orthogonal array to study the influence of each process parameters on the green machining parameters. Further, the effect of each AWJM parameters on the responses and surface quality of the machined surfaces are analyzed using parametric and SEM images, respectively. In addition, regression analysis, ANOVA and empirical models are performed to show the statistical significance of the green machining (i.e., AWJM) process and for optimal prediction of the green attribute of AWJM process, respectively. Finally, confirmatory tests are performed to verify results with experimental results.

3.2 Material and Method

In this work, WDFRP composite is prepared by using a natural filler such as Shagun wood dust (SWD). Sundi tree is available easily in the northeast region of India. Wood dust is collected when cutting Sundi tree wood. The main constituents of the Sundi wood are cellulose, glucomannan, xylem, and lignin [19]. SWD with particle size 400 μm and density 0.779 g/cc is seen in Fig. 3.1a. SWD is washed by using distilled water and placed in an electric oven for drying (Fig. 3.1b).

To prepare the workpiece samples, a dough that is epoxy filled with SWD was mechanically stirred and gradually poured into vacuum glass chamber [3]. Before that epoxy resin Araldite LY 556 having density 1.26 g/cm^3 was mixed with harder HY951 in the ratio 10:8 by weight. For moisture removal purposes, the mixer was cured for 24–48 h at room temperature. As shown in Fig. 3.1c, the prepared specimen having dimensions 180 mm × 140 mm × 6 mm was used for AWJM [3]. Figure 3.1d shows the AWJM CNC machine tool (DARDI International Corporation, China)

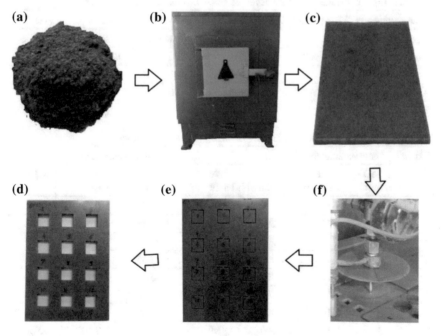

Fig. 3.1 **a** Sundi wood dust, **b** electric oven, **c** WDFRP composite specimen

used in the present work. The discharge rate of 2.31 l/min and an orifice diameter of 0.25 mm, designed pressure of 3800 bars, were taken during the experimentation [14]. A voltage of 150 V, current 20 A, and nozzle angle 90° were the major inputs. Abrasive material of type garnet with of size 210 μm, i.e., 70 mesh, 177 μm, i.e., 80 mesh, and 149 μm, i.e., 90 mesh, mixed with a distilled water at room temperature was used as an abrasive slurry [12]. Throughout the experiments, the working specimen (Fig. 3.1c) on size 180 mm × 140 mm × 6 mm was taken for machining.

The experiment is performed on WDFRP composite (Fig. 3.1e) by using an L_9 orthogonal array with the help of four input parameters such as SoD, WP, NS, and AGS are depicted in Table 3.1, and the final machined specimen is ready for the parametric analysis [14] as shown in Fig. 3.1f. Each experiment is performed with three times, and the average value of a response parameter is taken for analysis [20] as depicted in Table 3.2.

Following equations have been used to evaluate output parameters or responses:

$$MRR = \frac{W_i - W_f}{MT} \qquad (3.1)$$

where W_i and W_f are initial weight and the final weight; MT is the machining time in second; V is the voltage in volt; and I represent current supplied in amp; s of

Table 3.1 Input parameters and their levels of green machining (i.e., AWJM) process

Input parameters	Symbol	Units	Level 1	Level 2	Level 3
Standoff distance	SoD	mm	1	2	3
Work pressure	WP	MPa	100	125	150
Nozzle speed	NS	mm/min	100	200	300
Abrasive grain size	AGS	mesh	70	80	90

Table 3.2 Experimental results of green machining (i.e., AWJM) process

Exp. No.	Input parameters				Output parameters		
	SoD (mm)	WP (MPa)	NS (mm/min)	AGS (mesh)	MRR (g/mm)	SR (μm)	MT (s)
1	1	100	100	70	2.53	0.128	0.171
2	1	125	200	80	11.15	0.116	0.288
3	1	150	300	90	26.54	0.105	0.613
4	2	100	200	90	23.35	0.133	0.392
5	2	125	300	70	45.31	0.134	0.459
6	2	150	100	80	17.78	0.114	0.723
7	3	100	300	80	64.08	0.106	0.137
8	3	125	100	90	23.88	0.102	0.521
9	3	150	200	70	53.85	0.104	0.142

the workpiece; MRR is the material removal rate, and MT is the machining time. Thereafter, the machined surface was measured using surface profilometer (make: Tokyo Seimitsu Co. Ltd. Model: Handysurf E-35B) to determine the SR. During the analysis, the parameter MRR is taken as higher-the-better (HB) while SR and MT are set to be smaller-the-better (SB) [21]. The results of the experiment are provided in Table 3.2.

3.3 Results and Discussion

3.3.1 Parametric Analysis

3.3.1.1 Effect of Input Parameters on MRR

Figure 3.2a–d presents the variation of responses with AWJM parameters. From the mean effect plot, a similar pattern of MRR is observed for the parameter SoD and NS. In the case of SoD, MRR is increased (i.e., 13.40–47.27 g/mm) with increasing of SoD (i.e., 1–3 mm) as shown in Fig. 3.2a. Because at the increases of SoD, increases the jet flow radius of AWJM process and covers the larger area for the impact which

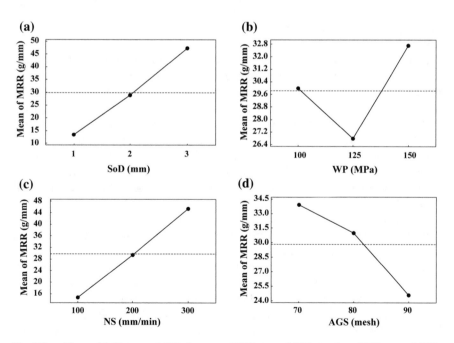

Fig. 3.2 a Mean of SoD versus MRR, **b** mean of WP versus MRR, **c** mean of NS versus MRR, **d** mean of AGS versus MRR

results in an increase of MRR as seen in Fig. 3.2a [7, 22, 23]. Similarly, in the case of NS, the rate of MRR is found to be same as SOD, i.e., increase of MRR from 14.73 to 45.31 g/mm when the increase of NS from 100 to 300 mm/min. This is due to the fact that the kinetic energy of the abrasive particles inside the nozzle is increased with increasing NS, which results in higher MRR [21] While in the case of WP, decrement is observed in the MRR (i.e., 29.98–26.78 g/mm) up to mid setting (i.e., 100–125 MPa) afterward MRR is increased from 26.78 to 32.72 g/mm, when WP setting is 125–150 MPa as shown in Fig. 3.2b. Because, at higher WP, the penetration of the abrasive particle is more due to higher force obtained by higher pressure and able to remove more material [21]. With reference to the parameter AGS as observed in Fig. 3.2d, MRR is decreased from 33.89 to 24.59 g/mm when AGS is increased from 70 to 90 mesh. The reason behind that, the smaller size of abrasive particles can easily penetrate in the WDFRP composite which helps to remove more material. But in the case of larger abrasive particles, lesser MRR is obtained because due to more surface area of the larger size abrasive particles cause restriction to penetrate the abrasive particles into the working specimen. This reason, lesser MRR is obtained in the case of larger AGS as can be seen in Fig. 3.2d [5, 13]. Based on parametric analysis of MRR, the combination of the best setting for higher MRR is SoD (3 mm, level 3), WP (150 MPa, level 3), NS (300 mm/min, level 3), and AGS (70 mesh, level 3).

3.3.1.2 Effect of Input Parameters on SR

Quality of the machined surface can be evaluated by the amount of surface irregularity, i.e., roughness. Appropriate combination of machining parameters can generate the desired surface roughness for better functional performance of surfaces. The effect of independent parameters such as SoD WP, NS, and AGS on SR is depicted using main effect plot seen in Fig. 3.3a–d. As per Fig. 3.3a, it is observed that the value of SR is increased from 0.117 to 0.127 μm as SoD is increased from 1 to 2 mm, and then a sudden decline in the magnitude of SR value from 0.127 to 0.104 μm when SoD is increased from 2 to 3 mm because, at higher SoD, the kinetic energy of the abrasive particles is more which helps to remove larger MRR with a lesser SR value [6].

On the other hand, lower SoD creates the lesser kinetic energy of the abrasive particles and lesser material removal from work specimen and rougher surface. Moreover, SR value is reduced from 0.122 to 0.108 μm with the increase of WP from 100 to 150 MPa as shown in Fig. 3.3c. It can be attributed to the fact that higher WP generates a more cutting force of abrasive particles results in the smoother cut as can be seen in Fig. 3.3c [24]. As per observation in Fig. 3.3b, the SR value increases as an increase of NS from 100 to 200 mm/min and then decreases from 200 to 300 mm/min. This is because, at lower NS, abrasive particles get more time to penetrate, whereas at higher NS, minimum time available for penetration, resulting in a decrease in SR value [15]. However, in the case of AGS, as shown in Fig. 3.3d, the SR value is decreased from 0.122 to 0.112 μm as an increase of AGS from 70 to 80 mesh. Afterward, a slight

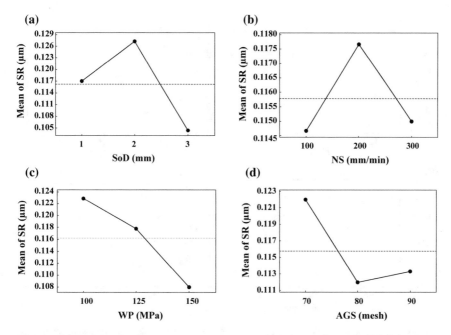

Fig. 3.3 **a** Mean of SR versus SoD, **b** mean of SR versus WP, **c** mean of SR versus NS, **d** mean of SR versus AGS

increment is observed in SR value (i.e., 0.112–0.113 μm) with an increase of AGS from 80 to 90 mesh. This is because for the larger size of abrasive particles which occupies more area of cutting width and able to remove more material on a single strike in WDFRP composite results in proper cut and lesser surface roughness [25]. The recommended optimal settings for SR based on the above analysis are SR is SoD (3 mm, level 3), WP (150 MPa, level 3), NS (100 mm/min, level 1), and AGS (80 mesh, level 2).

3.3.1.3 Effect of Input Parameters on MT

Figure 3.4a–d depicts the effect of AWJM parameters on MT. The increment in MT is observed from 0.350 to 0.518 min when SoD increases from 1 to 2 mm; afterward, MT is drastically decreased from 0.518 to 0.268 min as the increase of SoD is increased from 2 to 3 mm (Fig. 3.4a). This is because, at higher SoD, erosion of abrasive particles at the inner wall of the WDFRP composite is straight, results in lesser MT [25]. However, MT is increased from 0.231 to 0.490 min as an increase of WP from 100 to 150 MPa as shown in Fig. 3.4b. This is because, at higher WP, abrasive particle becomes fragmented and loses its cutting ability that results in higher MT [23].

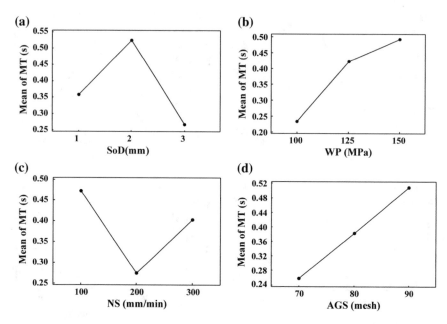

Fig. 3.4 **a** Mean of MT versus SoD, **b** mean of MT versus WP, **c** mean of MT versus NS, **d** mean of MT versus AGS

On the other hand, the drastic decrement of MT value (i.e., 0.469–0.269 min) is observed. At the time of NS is increased from 100 to 200 mm/min as well as the significant increment is observed in MT as an increase of NS value from 200 to 300 mm/min as observed in Fig. 3.4c. This is because, when NS increases, the kinetic energy of the abrasive particles at the tip of the nozzle is also an increase which helps to remove the material on jet flow direction, producing lesser MT. But at very high nozzle speed, the abrasive particles get shorter available time to penetrate, resulting in lesser material removal, and hence, the higher MT is produced [13]. As per observation in Fig. 3.4d, MT is increased from 0.251 to 0.504 s when AGS increases from 70 to 90 mesh. Due to the smaller size of AGS particles easily erodes, the inner wall of the hole and unwanted material is removed easily. Thus, lesser MT is produced at lower AGS [8]. With reference to Fig. 3.4a–d, the arrangement of the best setting for lesser MT is SoD (3 mm, level 3), WP (100 MPa, level 1), NS (200 mm/min, level 2), and AGS (70 mesh, level 1).

3.3.2 ANOVA Study

The statistical method, i.e., ANOVA is used to study the influence of each machining parameters of AWJM process on WDFRP composite. The analysis is carried out in Minitab 17 software, and their results are depicted in Tables 3.3, 3.4, and 3.5. In

Table 3.3 Analysis of variance for MRR

Source	DF	Adj. SS	Adj. MS	F-value	P-value
Regression	5	3316.63	663.327	2966.41	0.000
SoD	1	0.02	0.018	0.08	0.797
WP	1	0.02	0.018	0.08	0.797
NS	1	17.93	17.929	80.18	0.003
SoD × NS	1	181.02	181.016	809.51	0.000
WP × NS	1	17.85	17.853	79.84	0.003
Error	3	0.65	0.224		
Total	8	3317.31			

$R^2 = 99.98\%$, R^2 (Adj.) $= 99.95\%$, R^2 (Pred.) $= 99.39\%$

Table 3.4 Analysis of variance for SR

Source	DF	Adj. SS	Adj. MS	F-value	P-value
Regression	6	0.000984	0.000164	0.96	0.590
SoD	1	0.000452	0.000452	2.66	0.245
WP	1	0.000185	0.000185	1.09	0.406
NS	1	0.000247	0.000247	1.45	0.351
SoD × WP	1	0.000398	0.000398	2.34	0.049
SoD × NS	1	0.000139	0.000139	0.82	0.461
WP × NS	1	0.000223	0.000223	1.31	0.371
Error	2	0.000340	0.000170		
Total	8	0.001324			

$R^2 = 89.32\%$, R^2 (Adj.) $= 0.00\%$, R^2 (Pred.) $= 0.00\%$

Table 3.5 Analysis of variance for MT

Source	DF	Adj. SS	Adj. MS	F-value	P-value
Regression	3	0.1852	0.06174	1.74	0.274
SoD	1	0.1162	0.11624	3.28	0.13
NS	1	0.1193	0.11933	3.37	0.126
SoD × NS	1	0.1674	0.1674	4.73	0.042
Error	5	0.1771	0.03542		
Total	8	0.3623			

$R^2 = 85.12\%$, R^2 (Adj.) $= 82.80\%$, R^2 (Pred.) $= 0.00\%$

this analysis, if probability value, i.e., P-value is less than 0.05, shows the effective significance of the parameters, while larger F-value indicates that the variation of process parameters makes an extreme change in the performance. Also, the stepwise elimination method is used to remove insignificant parameters to adjust the fitted quadratic model [26].

The result of ANOVA for MRR is depicted in Table 3.3. The parameter WP × NS (interaction term) and NS (linear term) are found to be more significant for MRR and have the same P-value and F-Value, i.e., 0.003 and 80.18, respectively. However, the interaction parameter SoD × NS is the most influence for MRR because of highest F-value, i.e., 809.51 and zero probability value. Also, the R^2 value of MRR is 99.98%, which indicates that the presented model fit the data very well. The ANOVA result is listed in Table 3.5 for MT. The result shows that the interaction parameters, i.e., SoD × WP, are significant because the P-value is found to be less than 0.05, i.e., 0.049 and F-value is 2.34. While the value of R^2 obtained for MRR is 89.32%, which indicates that the presented model fits the data very well. The ANOVA result (Table 3.5) shows that the F-value for the interaction of SoD with NS is larger, i.e., 4.73 and P-value is less than 0.05, i.e., 0.042. It indicates that the interaction of SoD with NS is most significant for MT. The obtained R^2 value is 85.12%, which indicates from the analysis that the presented model fits the data.

3.3.3 Empirical Model

The empirical model is necessary to predict the performance of response parameters, and it is developed by using regression analysis. The regression equations for MRR, SR, and MT are listed in Eqs. (3.3–3.5), respectively.

$$MRR = -1.29 - 0.179 \times SoD + 0.0072 \times WP - 0.1786 \times NS$$
$$+ 0.09284 \times SoD \times NS + 0.001166 \times WP \times NS \qquad (3.2)$$

$$SR = 0.1913 - 0.0838\,SoD - 0.000888\,WP + 0.001003\,NS$$
$$+ 0.000739\,SoD \times WP - 0.000109\,SoD \times NS - 0.000006\,WP \times NS \quad (3.3)$$

$$MT = -0.285 + 0.368\,SoD + 0.00373\,NS - 0.002046\,SoD \times NS \qquad (3.4)$$

As per Eq. (3.2), it is observed that variable WP, interaction SoD with NS, and interaction WP with NS are a positive effect on MRR and SoD, NS is a negative effect on MRR. However, Eq. (3.3) indicates that parameter NS and interaction parameter SoD with NS are positive effects on SR and parameter SoD, WP, interaction parameter SoD with NS and WP with NS are a negative effect on SR as well as with reference of Eq. (3.4) SoD and NS are positive effect, and interaction of SoD with NS is a negative effect. By observing Eqs. (3.2–3.4), parameter WP and interaction parameter SoD

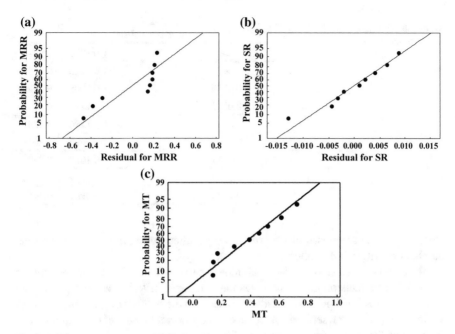

Fig. 3.5 **a** Probability plot for MRR, **b** probability plot for SR, **c** probability plot for MT

with NS, WP with NS are significant for MRR. Also, parameter NS and interaction parameter SoD with WP are important in the case of SR. Along with, SoD and NS are influencing parameters for MT. Similarly, Fig. 3.5a–c shows the probability plots for which all responses such as MRR, SR, and MT are closer to the reference line. Hence, it is observed that all experimental data are in close approximation with the reference line and shows the result that is satisfactory [26].

3.3.4 Modeling and Optimization

Optimization of AWJM process parameters for the improved machining of green composites is done using MOORA method. For the optimization, MRR, SR, and MT are considered as response/output parameters, whereas input AWJM parameters are as alternatives. In the MOORA method, first, the development of the decision matrix is done using Eq. (3.5).

$$X = \begin{bmatrix} X_{11} & X_{12} & \dots & X_{1n} \\ X_{21} & X_{22} & \dots & X_{2n} \\ \dots & \dots & \dots & \dots \\ X_{m1} & X_{m2} & \dots & X_{mn} \end{bmatrix} \tag{3.5}$$

Table 3.6 Normalized decision matrix (N_{ij})

Exp. No.	Material removal rate (MRR)	SR	MT
1	0.0094	0.1228	0.0496
2	0.0415	0.1113	0.0836
3	0.0989	0.1008	0.1779
4	0.0870	0.1276	0.1138
5	0.1688	0.1286	0.1332
6	0.0662	0.1094	0.2098
7	0.2387	0.1017	0.0398
8	0.0889	0.0979	0.1512
9	0.2006	0.0998	0.0412

where X_{ij} is the responses of ith criteria on jth alternatives, m and n are the total numbers of criteria's and alternatives, respectively.

The results of decision matrix for all nine experimental combinations are given in Table 3.6. The decision matrix includes the number of criteria's as response/output parameters of AWJM process and alternatives as to the number of process parameter setting or experimental settings. In the second step, normalization of the response data for each of the output parameters is performed using Eq. (3.6). The normalization process is required in order to convert the different unit of the responses into the comparable unit.

$$N_{ij} = \frac{X_{ij}}{\left[\sum_{i=1}^{m} x_{ij}^2\right]^{1/2}} \quad \text{where } j = 1, 2, \ldots n \tag{3.6}$$

where N_{ij} represents normalized performance values of ith criteria's corresponding to the jth alterantives.

Thereafter, determination of overall assessment of the criteria's/responses corresponding to each of the input parameter setting using Eq. (3.7) and the results are shown in Table 3.7. Finally, the ranking of the alternatives is done based on the overall assessment of the criteria's/responses and the results of the same are depicted in Table 3.6.

$$y_j = \sum_{i=1}^{g} N_{ij} - \sum_{i=g+1}^{n} N_{ij} \tag{3.7}$$

where g is the number of criteria's to be maximized, $(n - g)$ is the number of criteria's to be minimized, y_j denotes assessment values of ith criteria's with respect to the all jth alternatives.

It is observed that Exp. No. 5 shows highest assessment values among the other and found the optimal setting. The optimal setting obtained is SoD (2 mm, level 2), WP (125 MPa, level 1), NS (300 mm/min, level 3), and AGS (70 mesh, level 1).

Table 3.7 Assessment values (y_i) values

Exp. No.	Assessment values (y_i) value	Rank
1	0.1819	9
2	0.2364	8
3	0.3775	5
4	0.3284	7
5	**0.4306**	**1**
6	0.3854	3
7	0.3802	4
8	0.3380	2
9	0.3416	6

The optimal setting, i.e., Exp. 5 provides the most optimal responses which reduce the processing time, increases the MRR, better surface finish, and lesser wastes during the machining of WDFRP composites which directly or indirectly increases the efficiency and performance of the green machining process.

3.3.5 Confirmation Experiments

The confirmation test with prediction is depicted in Table 3.8. An optimal parameter setting such as SoD, WP, NS, AGS, and corresponding response parameters such as MRR, SR, **MT** is evaluated. The confirmatory test is performed to verify the optimal result obtained by parametric analysis. Also, predictions of each response parameter of AWJM process are determined. The result of the confirmatory test shows that the prediction found to be comparable.

3.3.6 Surface Integrity of Machined Surfaces

Additionally, magnified views of SEM images of the machined surface of WDFRP composites are analyzed to study surface integrity as shown in Fig. 3.6. The model ZEISS EVO-Series SEM EVO 18 manufactured by ZEISS machine is used for the extraction of SEM images. The machined surface obtained at the optimal setting using MOORA method is considered for the SEM images.

It has been observed from the SEM images Fig. 3.6a–e that, a contour of crack propagation was found on the machined surface. Definitely, it can be expected to happen in WDFRP composites, because of the varying degree of interfacial adhesion between the fillers and matrix materials. In the SEM images, only filler damages were observed instead of interlinear delamination as shown in Fig. 3.6b.

Table 3.8 Confirmation result

Confirmatory test results

Output parameters	Input parameters				Exp. results	Optimum parameters				Predicted exp. result	Confirmatory exp. result
	SoD (mm)	WP (MPa)	NS (mm/min)	AGS (mesh)		SoD (mm)	WP (MPa)	NS (mm/min)	AGS (mesh)		
MRR	3	100	300	80	64.080	3	150	300	70	67.612	67.581
SR	3	125	100	90	0.102	3	150	100	80	0.098	0.099
MT	3	100	300	80	0.13	3	100	200	70	0.11	0.112
OAV					0.430	2	125	300	75	0.412	0.421

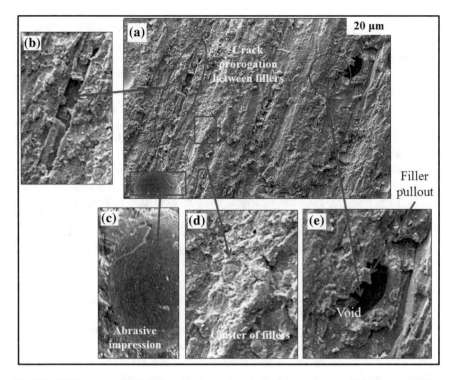

Fig. 3.6 SEM images of WDFRP composite in green machining process at (SoD = 2 mm, WP = 125 MPa, NS = 300 mm/min, AGS = 75 mesh) **a** machined surface, **b** crack in WDFRP composite, **c** abrasive impression, **d** cluster of fillers, and **e** void and filler pull out

On the other hand, some voids (Fig. 3.6e) were observed in some places of the machined surfaces due to the presence of moisture in natural fillers. Later voids prorate the crack in the composites when the stress is developed during the machining. Similarly, in some places, the cluster of fillers and abrasive impression were observed as shown in Fig. 3.6c–d.

3.4 Summary

This chapter presents the machining performance of green machining process on polymer composites. Abrasive water jet machining (AWJM), a commonly known as green machining process and wood filler-based polymer composites or green composites, is considered in this study. Taguchi L_{09} orthogonal array experiments are conducted to study the most influencing variables (SoD, WP, NS, and AGS) on the responses (MRR, SR, MT). Based on the experimental results, parametric analysis, regression analysis and optimization following conclusions are drawn from the present work:

- From the parametric analysis, optimal settings for MRR are SoD (3 mm, level 3), WP (150 MPa, level 3), NS (300 mm/min, level 3), and AGS (70 mesh, level 3); for SR are SoD (3 mm, level 3), WP (150 MPa, level 3), NS (100 mm/min, level 1), and AGS (80 mesh, level 2); and for MT are SoD (3 mm, level 3), WP (100 MPa, level 1), NS (200 mm/min, level 2), and AGS (70 mesh, level 1) are obtained.
- From the ANOVA, the parameters NS (linear term), SOD $\times$ NS (interaction term), WP $\times$ NS (interaction term) for MRR, the parameters SoD $\times$ WP (interaction term) for SR and SoD $\times$ NS (interaction term) for MT are found to be the most significant parameters.
- From the optimization: Exp. No. 5 yields the highest OAV among the other experimental setup and provides overall optimal setting, i.e., SoD (2 mm, level 2), WP (125 MPa, level 1), NS (300 mm/min, level 3), and AGS (70 mesh, level 1) is obtained.
- The optimal setting provides optimal responses such as the higher MRR, lesser MT, and better SR, which have less influence on a generation of environmental issues aroused during the machining of WDFRP composites in AWJM process.
- Additionally, prediction models are developed for MRR, SR, and MT for optimal prediction of AWJM responses. The result shows that predicted responses are close and satisfactory with experimental results.
- A significant amount of crack propagation, some voids on the machined surface were found due to the presence of moisture in the natural fillers.
- At last, a confirmatory test is performed to verify the experimental results. The result shows that confirmatory results and experimental results are comparable.

Finally, it is concluded that the AWJM process is capable to machine WDFRP composites under green machining environment. The optimal set of parameters as obtained in this research can be used as a ready industrial reference. Also, the developed empirical models for AWJM responses can be used for future predictions.

References

1. A.M. Al-Bakri, J. Liyana, M.N. Norazian, Mechanical properties of polymer composites with sugarcane bagasse filler. Adv. Mater. Res. **740**, 739–744 (2013)
2. M.K. Pradhan, Estimating the effect of process parameters on MRR, TWR and radial overcut of EDMed AISI D2 tool steel by RSM and GRA coupled with PCA. Int. J. Adv. Manuf. Technol. **68**, 591–605 (2013)
3. Bhowmik S. Jagadish, A. Ray, Prediction and optimization of process parameters of green composites in AWJM process using response surface methodology. Int. J. Adv. Manuf. Technol. **87**(5), 1359–1370 (2016)
4. M.M. Korat, G.D. Acharya, A review on current research and development in abrasive waterjet machining. Int. J. Eng. Res. Appl. **4**, 423–432 (2013)
5. R.V. Rao, V. Kalyankar, Optimization of modern machining processes using advanced optimization techniques: a review. Int. J. Adv. Manuf. Technol. **73**(5–8), 1159–1188 (2014)
6. V.H.C. Albuquerque, J.F.S. De-Marques, O.N.G. Andrade, Drilling damage in composite material. Materials **7**(5), 3802–3819 (2014)

7. J. Ahmad, *Machining of Polymer Matrix Composites* (Springer USA, Boston, MA, 2009), pp. 150–230
8. Jagadish, S. Bhowmik, A. Ray, Prediction of surface roughness quality of green abrasive water jet machining: a soft computing approach. J. Intell. Manuf. 1–15 (2016)
9. M.M. Korat, G.D. Acharya, A review on current research and development in abrasive waterjet machining. J. Eng. Res. Appl. **4**, 423–432 (2014)
10. J. Wang, Machinability study of polymer matrix composites using abrasive waterjet cutting technology. J. Mater. Process. Technol. **94**, 30–35 (1999)
11. N. Kazemi, Use of recycled plastics in wood plastic composites—a review. Waste Manage. **3**, 1898–1905 (2013)
12. K. Oksman, J.F. Selin, Plastics and composites from polylactic acid, in *Natural Fibers, Plastics and Composites* (Springer US, Boston, MA, 2004), pp. 149–165
13. A.A. Shaikh, P.S. Jain, Experimental study of various technologies for cutting polymer. Int. J. Adv. Eng. Technol. **2**(1), 81–88 (2009)
14. E. Lemma, L. Chen, E. Siores, J. Wang, Study of cutting fiber-reinforced composites by using abrasive water-jet with cutting head oscillation. Compos. Struct. **57**, 297–303 (2002)
15. D.K. Shanmugam, F.L. Chen, E. Siores, M. Brandt, Comparative study of jetting machining technologies over laser machining technology for cutting composite materials. Compos. Struct. **57**, 289–296 (2002)
16. J.H. Ke, F.C. Tsai, J.C. Hung, T.Y. Yang, B.H. Yan, Scrap wafer regeneration by precise abrasive jet machining with novel composite abrasive for design of experiments. Proc. IMechE Part B J. Eng. Manuf. **225**(6), 954–965 (2011)
17. K.S.K. Sasikumar, K.P. Arulshri, K. Ponappa, M. Uthayakumar, A study on kerf characteristics of hybrid aluminium 7075 metal matrix composites machined using abrasive water jet machining technology. Proc. IMechE Part B J. Eng. Manuf. **37**(1), 554–565 (2016)
18. S. Kalirasu, N. Rajini, J.T.W. Jappes, Mechanical and machining performance of glass and coconut sheath fibre polyester composites using AWJM. J. Reinf. Plast. Compos. **34**(7), 564–580 (2015)
19. K. Rahul, K. Kaushik, Optimization of mechanical properties of epoxy based wood dust reinforced green composite using Taguchi method. Int. Conf. Adv. Manuf. Mater. Eng. **5**, 688–696 (2014)
20. Z. Yue, C. Huang, H. Zhu, Optimization of machining parameters in the abrasive waterjet turning of alumina ceramic based on the response surface methodology. Int. J. Adv. Manuf. Technol. **71**, 2107–2114 (2014)
21. M. Mhamunkar, Optimization of process parameter of CNC abrasive water jet machine for titanium Ti 6Al 4V material. Int. J. Adv. Res. Sci. Eng. Technol. **3**, 1640–1646 (2016)
22. W. Konig, C. Wulf, P. Gral, H. Willerscheid, Keynote—papers machining of fibre reinforced plastics. CIRP Ann. Manuf. Technol. **34**, 537–548 (1985)
23. A. Sabur, M.Y. Ali, M.A. Maleque, A.A. Khan, Investigation of material removal characteristics in EDM of nonconductive ZrO_2 ceramic. Procedia Eng. **56**, 696–701 (2013)
24. S. Shanmugha, Influence of abrasive water jet machining parameters on the surface roughness of eutectic Al-Si alloy–graphite composites. Mater. Phys. Mech. **9**, 1–8 (2014)
25. P.P. Badgujar, M.G. Rathi, Taguchi method implementation in abrasive waterjet machining process optimization. Int. J. Eng. Adv. Technol. **3**, 66–70 (2014)
26. M.A. Azmir, A.K. Ahsan, A. Rahmah, Effect of abrasive water jet machining parameters on aramid fibre reinforced plastics composite. Int. J. Mater. Form. **2**, 37–44 (2009)

Chapter 4
Abrasive Water Jet Machining of Ceramic Composites

4.1 Introduction

Increased use of ceramic materials in aerospace, automobile, shipbuilding, and dental applications with different contour parts, viz. economical manufacturing route has paid significant attention toward researchers in the past two decades [1–3]. Poses excellent hardness, and strength even at elevated temperatures, low wear and corrosion resistance, electromagnetic and biocompatibility are the unique features of ceramic materials [4]. These features are often difficult or quite impractical for manufacturing complex parts economically, viz. traditional machining process. Traditional machining processing route of ceramic materials using high-speed milling followed by cemented carbide tools to complete parts to final geometry is often problematic and time-consuming [5]. This occurs due to high material strength at elevated temperatures and offers greater resistance for the part geometry to be machined. Higher resistance offered for ceramic machining might be due to the involved deformation mechanism especially at lower feed rate and shallow depth of cut which generates larger tool wear as a result of higher cutting forces [6]. New and emerging technologies are in great demand which offer a single-step machining of ceramic materials to complex geometries that limit the said disadvantages.

In recent years, non-traditional cutting technology showed greater potential to shape complex geometry parts for difficult-to-cut materials including high-strength ceramics [7]. Non-traditional machining processes, namely EDM, WEDM, laser beam machining (LBM), and electron beam machining (EBM), etc., are available for machining of difficult-to-machine materials [8]. Machining of ceramic using EDM or WEDM is not applicable [9] due to brittle and low hardness. EDM also cut parts to final geometry but require post-processing due to the recast layer formation and induced tensile stresses due to the HAZ. Machining speed is approximately similar for both EDM and AWJM, and however, EDM produces coarser surface in addition to thickness limitations [5]. Despite its advantages, EDM or WEDM process is considered to be a hazardous process because it discharges a large number of toxic components which are in the forms of solid, liquid, and gaseous waste during the machining of ceramic materials that results in serious health and environmental issues. In the case of a laser machining (LM) of ceramic composites, damage as well as burning of composites is due to large heat-affected zone [10]. Moreover, USM process can be utilized for machining of ceramic composites. But, USM process needs higher hardness materials with good strength and there are thickness limitations with this process. As ceramic composites are generally of lower hardness, USM is not appropriate for machining of ceramic composites [11].

Besides the advantages of non-conventional for machining of ceramics, these are considered to be a hazardous process because it discharges a large number of toxic substances in the form of solid, liquid, and gaseous waste, during the machining resulting in serious occupational health and environmental issues [12]. In order to overcome aforementioned issues, AWJM process is used for machining of composite materials. As discussed in the previous chapters, AWJM process is capable of machining of all kinds of engineering materials including ceramics. It offers many benefits such as reduced waste generation, no thermal distortion due to cold cutting mechanism, low cutting forces which do not cause chatter, higher machining flexibility and versatility, less environmental contamination, less sensitive to alter material properties, do not generates ant forms of dust or aerosols or fumes, hence, sometime AWJM process is known as environment friendly or green machining or green manufacturing process [13, 14].

In the past, various researchers worked on the machining of ceramic composite using non-conventional processes, but no work is reported on machining of ceramic specifically zirconia (ZrO_2) composite using green machining process, i.e., AWJM process.

The main objective of this research work is to study the experimental investigation on the machining performance of green machining process during the machining of zirconia (ZrO_2) composite. For this, an environmental friendly machining process or green machining process known as abrasive water jet machining (AWJM) is used for machining of zirconia (ZrO_2) composite. Experiments are conducted using the Taguchi (L_{27}) method to analyze the influence of AWJM parameters (standoff distance, working pressure, and nozzle speed) on MRR, surface roughness, and process energy. Further, the effect of each AWJM parameters on the responses and surface quality of the machined surfaces are analyzed using parametric and SEM images, respectively. In addition, empirical models are developed for optimum prediction of responses. Finally, confirmatory tests are performed to verify results with experimental results.

Fig. 4.1 Zirconia (ZrO_2) composite

4.2 Material and Method

4.2.1 Material Specimen

In the present work, the work specimen (zirconia (ZrO_2) composite) has been prepared by a sol-gel method having a bulk density of 6 g/cc and is stabilized with 8 mol% of yttrium (yttrium oxide, Y_2O_3). The dimensions of the workpiece are shown in Fig. 4.1. The workpiece is square shaped with each side having a length of 100 mm and a thickness of 3 mm. The machining is done for a through hole of depth 3 mm.

4.2.2 Taguchi Method

The Taguchi robust method employs a well-defined procedure that starts with selecting an experimental design with different levels and collects output data that offer a precise estimate of factor (i.e., main and interaction) effects [15]. In addition, the collected data were utilized to minimize the impact of noise factors and determine the optimal set of factors that could maximize or minimize the response function. The cost-effective Taguchi method saves considerable efforts, time, and resources to achieve the aforementioned task with few experimental trials as discussed below,

$$\text{DOF} = (L-1)V + (L-1)I + 1 \tag{4.1}$$

where the term DOF represents degrees of freedom, V is the number of independent factors, L depicts the number of levels, and I defines the number of interaction terms. In the present work, $V = L = 3$ and $I = 3$, and hence, the total DOF $= 19$. L_{27} orthogonal array was chosen for modeling, analysis, and optimization of green machining process i.e. AWJM process.

4.2.3 Experimental Procedure

The CNC Water Jet Cutting Machine manufactured by DARDI International Corporation, China was used for experimentation is shown in Fig. 4.2. The designed pressure of 3800 bar, discharge rate as 2.31 l/min, and an orifice diameter of 0.25 mm were taken during the experimentation. Abrasive material of type Garnet with of size (70 [$\approx$ 210 µm]) mixed with a distilled water at room temperature was used as an abrasive slurry. Throughout the experiments, the voltage of 300 V, a current of 20 A, and nozzle angle of 900 were input to the AWJ machine. The work specimen of size 100 mm $\times$ 100 mm $\times$ 3 mm was taken for machining.

The selected L_{27} orthogonal arrays representing the different combination of experiments are conducted by using the experimental setup shown in Fig. 4.2 and input variable their levels as tabulated in Table 4.1. Results of some pilot experiments and machine constraints are considered to decide the fixed parameters during experimentation (refer to Table 4.2).

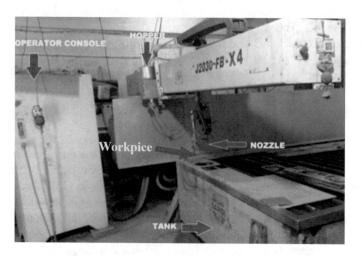

Fig. 4.2 AWJM experimental setup

Table 4.1 AWJM operating variables and levels

Control variables	Notation	Levels (1, 2, 3)
Standoff distance, mm	A	1.5, 2.5 and 3.5
Working pressure, MPa	B	100, 125 and 150
Nozzle speed, mm/min	C	100, 200 and 300

Table 4.2 Fixed parameters of AWJM

Abrasive	Garnet
Abrasive grit size	70 mesh ($\approx$210 μm)
Abrasive particle shape	Angular
Orifice diameter	0.25 mm
Focussing tube diameter	1.02 mm
Focusing length	76 mm
Current and voltage	20 A and 300 V
Material dimension	100 mm × 100 mm × 3 mm
AWJM system	Pressure intensifier, injection type nozzle
Jet impact angle	90°
Nozzle material	Carbide
Work material	Ceramics
Density of material	

The Taguchi robust technique is employed to design the experiments for three factors set at three levels (Table 4.1). Accordingly, L_{27} OA-based experimental design is selected which provides a precise estimation of both individual and interaction factor effects. The L_{27} OA matrix used to perform experiments is presented in Table 4.3. For each experimental set, the experiments are repeated for three times and the mean values of the measured MRR and SR are treated for modeling, analysis, and optimization. MRR is determined for each set of experimental conditions by weighing the work material using precise digital weighing balance possessing an accuracy of 1 mg. During experimentation, the initial and final weights of the workpiece are recorded and the obtained values are used to estimate the material removal rate using Eq. (4.2).

$$\text{MRR} = \frac{(\text{Initial weight of work piece} - \text{Final weight of work piece} - \text{Weight of cut specimen})}{\text{Machining time}}$$

$$= \frac{(w_i - w_f - w_c)}{t_m} = \text{mg}/\min \tag{4.2}$$

Similarly, the obtained machined surface for each experimental condition was measured by utilizing Surfcom 130A-Monochrome (make: Tokyo Seimitsu Co. Ltd). As stated earlier, for industry and customer viewpoint, the objective function should be higher-the-better (HB) for MRR and lower-the-better (SB) for SR. The computation of signal-to-noise (S/N) ratio for both the objective functions is expressed as discussed below,

$$S/N_{NB} = -10 \log_{10} \left(\frac{1}{n} \sum_{i=1}^{n} \frac{1}{y_i^2} \right) \tag{4.3}$$

Table 4.3 Combinations of AWJM parameters based on L_{27} orthogonal array and experimental results for average SR and MRR

Exp. No.	SOD, A	WP, B	NS, C	MRR, mm³/min	SR, μm	S/N of MRR	S/N of SR
1	1.5	100	100	235	0.189	47.42	14.48
2	1.5	100	200	226	0.201	47.08	13.94
3	1.5	100	300	208	0.211	46.36	13.51
4	1.5	125	100	394	0.168	51.91	15.50
5	1.5	125	200	381	0.178	51.62	14.99
6	1.5	125	300	391	0.178	51.84	14.99
7	1.5	150	100	651	0.249	56.27	12.08
8	1.5	150	200	586	0.213	55.36	13.43
9	1.5	150	300	343	0.258	50.71	11.77
10	2.5	100	100	169	0.272	44.56	11.31
11	2.5	100	200	314	0.267	49.94	11.47
12	2.5	100	300	397	0.273	51.98	11.28
13	2.5	125	100	583	0.291	55.31	10.72
14	2.5	125	200	492	0.281	53.84	11.03
15	2.5	125	300	381	0.283	51.62	10.97
16	2.5	150	100	726	0.331	57.22	9.60
17	2.5	150	200	804	0.334	58.11	9.53
18	2.5	150	300	610	0.334	55.71	9.53
19	3.5	100	100	382	0.391	51.64	8.16
20	3.5	100	200	475	0.372	53.53	8.59
21	3.5	100	300	245	0.371	47.78	8.62
22	3.5	125	100	480	0.349	53.63	9.14
23	3.5	125	200	602	0.344	55.59	9.27
24	3.5	125	300	397	0.354	51.98	9.02
25	3.5	150	100	916	0.371	59.24	8.62
26	3.5	150	200	853	0.359	58.62	8.89
27	3.5	150	300	602	0.354	55.60	9.02

$$S/N_{\mathrm{NB}} = -10 \log_{10} \left(\frac{1}{n} \sum_{i=1}^{n} y_i^2 \right) \qquad (4.4)$$

where the term y_i represents the actual experimental response data at ith trial and n depicts the number of experimental runs. The input–output data representing machining process parameters, i.e., AWJM parameters (NS, SOD, and WP) and green machining quality characteristics (MRR and SR) of L_{27} OA experiments are presented in Table 4.3.

4.3 Results and Discussion

Taguchi L_{27} OA experiments representing a different set of control variables have been used to conduct experiments and collect output data (MRR and SR). Statistical tests on the experimental input–output data were performed to examine the factors (i.e., linear and interaction) contribution, surface plots examine the relationship behavior of outputs with inputs and derive response equations. Furthermore, multiple objective functions are optimized for a single set of input conditions utilizing MOPSO-CD. The detailed analysis has been discussed as follows.

4.3.1 Parametric Analysis

4.3.1.1 Effect of Control Variables on MRR and SR

To optimize the process accurately, the nature of inputs on outputs are essential. Main effect plots determine the nature (behavior) of control parameter on output studied at various levels. Figure 4.3 depicts the response graphs which detail the variation of individual control factor studied at different levels on quality characteristics of AWJM process. MRR showed an increasing trend with an increase in WP and SOD (refer Fig. 4.3a). This occurs due to the machining area subjected by the impact of water increases with increased standoff distance, wherein the kinetic energy of abrasive particles impinge on the work surface also increases with higher work pressure leads to higher MRR. The necessary time for the abrasive particles to cut the work material decreases and jet deflection increases at higher NS. It resulted in low values of MRR. The optimal levels for each parameter are selected according to the determined average value of S/N ratio (refer Table 4.4). The optimal level thus selected for each parameter corresponds to the highest value of the computed S/N ratio for both MRR and SR (refer Fig. 4.3a, b). $A_3B_3C_2$ and $A_1B_2C_2$ are the optimal levels for the set of parameters correspond to MRR and SR.

Furthermore, working pressure contributes more followed by standoff distance and nozzle speed for MRR (refer Table 4.4). The combined influence of higher work pressure and standoff distance increases the driving forces of abrasive particles to hit the work surface at higher impact which develops craters of more depth results in more MRR. Contrary, the standoff distance showed maximum impact followed by working pressure and nozzle speed to reduce surface roughness. Lower the standoff distance, lesser will be the distance travelled by the abrasive particle which retains its sharp cutting ability as a result of avoiding intercollision between particles results in reduced SR. In addition, the development of large craters as a result of the momentum of the impact of abrasive particles on the work surface is less at higher work pressure that leads to reduced SR. Note that, NS is found to have negligible importance on the SR.

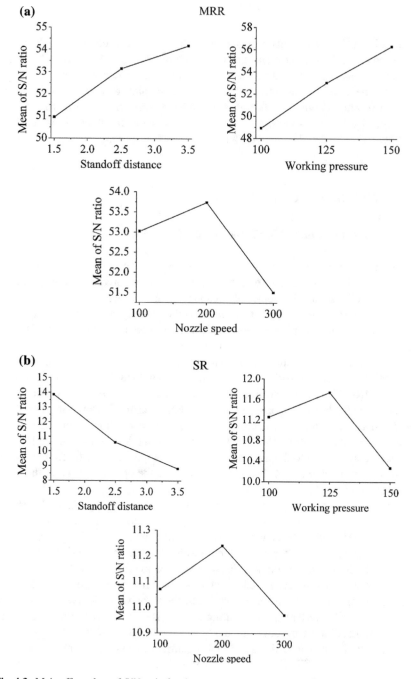

Fig. 4.3 Main effect plots of *S/N* ratio for the response: **a** MRR and **b** SR

Table 4.4 Mean *S/N* ratio response for MRR and SR

Levels (L)	Material removal rate			Surface roughness		
	A	B	C	A	B	C
L − 1	50.95	48.92	53.02	**13.85**	11.26	11.07
L − 2	53.14	53.04	**53.74**	10.60	**11.74**	**11.24**
L − 3	**54.18**	**56.32**	51.51	8.81	10.27	10.97
Max. − Min.	3.23	7.39	2.23	5.04	1.46	0.27
Rank	2	1	3	1	2	3

Bold values indicates: *Optimal Values

4.3.1.2 Surface Plot Analysis for MRR

The 3D graphical representation of the plot explains the geometric nature (maxima, minima, linear, and nonlinear) of the response surface when examined to know the individual and cumulative effective of the variables. The following key observations are drawn from the surface plot analysis of MRR. It is seen that, Fig. 4.4a displays the behavior of MRR represented by varying simultaneously the standoff distance and working pressure, after holding the nozzle speed at 200 mm/min. It is clear from the response surface that, MRR increases linearly with a cumulative increase of standoff distance and working pressure. The observation cleared their lies a negligible impact of standoff distance compared to that of working pressure.

An increase in standoff distance would increase the material removal rate linearly, whereas the material removal rate is seen to decrease gradually with increase in nozzle speed (refer Fig. 4.4b). The resulting response surface is seen to be almost flat for nozzle speed, which indicates a negligible impact on MRR. Figure 4.4c depicts there would be a rapid increase in material removal rate with the increase in working pressure, and MRR decreases linearly with an increase in nozzle speed. The impact of the cumulative effect of working pressure and nozzle speed is more compared to other interactions (refer Fig. 4.4a–c). Increased standoff distance allows the jet to expand which enhance the machining area coupled with increased kinetic momentum of abrasive particles striking the work surface with higher working pressure resulted in more material removal (refer Fig. 4.4a–c). As the nozzle speed increases, the total number of abrasive particles allowed to strike on the machining (i.e., target) area decreases which results in low MRR (refer Fig. 4.4b–c).

4.3.1.3 Surface Plot Analysis of SR

Figure 4.5 presents the variation of surface roughness with process parameters. It is clear from the surface plots that the desired minimum surface roughness lies close to the low values of standoff distance, working pressure, and nozzle speed. However, Fig. 4.5 shows reduced surface roughness could be the result of low values

Fig. 4.4 3D surface plots of MRR with **a** standoff distance (*A*) and working pressure (*B*), **b** standoff distance (*A*) and nozzle speed (*C*), and **c** working pressure (*B*) and nozzle speed (*C*)

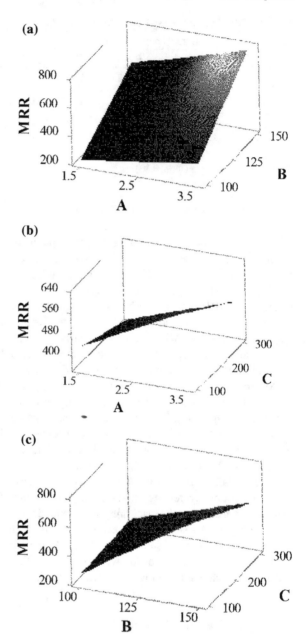

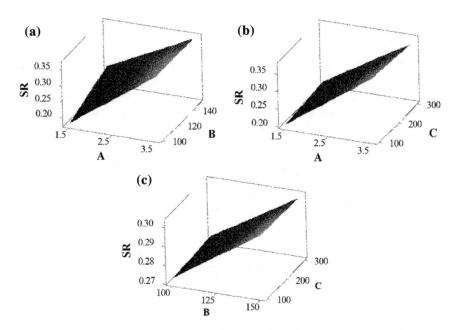

Fig. 4.5 3D surface plots of SR with **a** standoff distance (*A*) and working pressure (*B*), **b** standoff distance (*A*) and nozzle speed (*C*), and **c** working pressure (*B*) and nozzle speed (*C*)

of standoff distance, after maintaining the nozzle speed and working pressure kept at fixed middle levels. The optimal levels ($A_1B_2C_2$) for the reduced surface roughness seen to be slightly contradictory might be due to the dominant effect of standoff distance compared to working pressure and nozzle speed (refer Table 4.4). Figure 4.5 shows the surface roughness seen to have a linear effect with standoff distance, whereas it behaves nonlinearly with working pressure and nozzle speed. Increase in standoff distance increases the distance to be travelled by the abrasive particles which cause reduced cutting ability due to loss of sharpened cutting edges as a result of intercollision among the particles. Therefore, lower standoff distance generates a smooth surface as a result of improved kinetic energy (refer Fig. 4.5a). The response surface of SR with working pressure is seen to be almost flat due to the dominant impact with standoff distance. Increase in working pressure offers a sufficient amount of energy supplied by the abrasives without causing radical nozzle deflection results in steady waviness in surface roughness (refer Fig. 4.5a). Although the impact of nozzle speed is less toward surface roughness (refer Fig. 4.5b, c), SR increases with increase in NS due to few abrasive particles with less overlapping cutting action.

4.3.2   Empirical Model

The models are developed based on experimental data for both MRR and SR. Minitab 17 software platform conducts regression analysis to know the impact of control variables on output data and derives mathematical regression equations. Analysis of variance (ANOVA) tests the practical importance in terms of significance or insignificance when tested at a preset confidence level set at 95%. 3D surface plots explain the projection or predict the behavior (linear or nonlinear) of response under control variable constraints. The derived empirical relationship relating output expressed mathematically as a function of control variables (refer Eqs. 4.3 and 4.4).

$$MRR = -964 - 31A + 10.01B + 3.22C + 1.193AB - 0.163AC - 0.02673BC$$

(4.5)

$$SR = -0.194 + 0.1615A + 0.002031B + 0.000188C - 0.000563AB$$
$$- 0.000061AC - 0.0000001BC$$

(4.6)

4.3.3   ANOVA Study

Statistical tests are performed to know the contribution of both individual and combined factors' effects by using ANOVA. The significance of the said factors is tested subjected to the preset 95% confidence level. The adequacy of the developed models for both outputs is found to be statistically adequate as they produced a good coefficient of correlation (i.e., R close to 1) found equal to 0.9180 for SR and 0.866 for MRR (refer Table 4.5), respectively.

All linear (i.e., A, B, C) and corresponding interaction terms are found significant for MRR (refer Table 4.6). Few terms (i.e., AB and AC) in the fitted models are insignificant due their obtained P-value found to be greater than 0.05 (Table 4.5). This resulted in the lowest percent contribution by the terms, i.e., AB and AC (refer Fig. 4.6). Interesting to note that, although the individual parameters (i.e., A and B) posses greater percent contribution, their interaction term (i.e., AB) is found insignificant for MRR.

Table 4.5 Multiple correlation coefficient and insignificant terms

Output	Correlation coefficient		Terms	
	All R terms	Insignificant terms	Significant (P-value < 0.95)	Insignificant (P-value > 0.95)
SR	0.9180	0.8933	A, B, and AB	C, AC, and BC
MRR	0.8666	0.8266	A, B, C and BC	AC and AB

Table 4.6 ANOVA for MRR and SR

Output		Material removal rate, mm^3/min				Surface roughness, μm			
Details	DF	Adj. SS	Adj. MS	F	P	Adj. SS	Adj. MS	F	P
Model	6	907561	151260	21.66	0.000	0.118494	0.019749	37.30	0.000
Linear	3	840078	280026	40.10	0.000	0.115664	0.038555	72.81	0.000
Interaction	3	67482	22494	3.22	0.045	0.002830	0.000943	1.78	0.183
Error	20	139669	6983			0.010590	0.000530		
Total	26	1047230				0.129084			

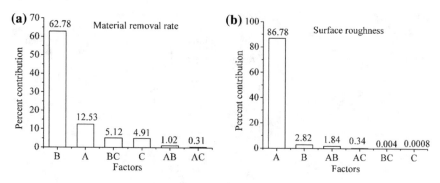

Fig. 4.6 Pareto ANOVA graph for **a** MRR and **b** SR

All linear terms (except nozzle speed) are statistically adequate for the preset 95% confidence level for SR. However, the combined effect of all interaction terms is found insignificant tested under the preset confidence level of 95%. However, *AB* and *AC* interaction terms in the fitted models are statistically insignificant wherein their corresponding *P*-value > 0.5. This resulted in the said interaction terms produced the lowest percent contribution toward SR. Although insignificant term contributions are less, they need not be removed from the fitted models as they result in an imprecise input–output relationship and reduce prediction accuracy [16]. Standoff distance followed by working pressure and their combined effects is statistically significant toward SR.

4.3.4 Modeling and Optimization

Particle swarm optimization (PSO) uses computational swarm intelligence-based evolutionary technique to optimize the parameters for multimodal responses of AWJM process. In 1992, John Kennedy was first credited for the development of PSO to solve the complex real-world problems [17]. PSO uses the basic underlying principle which mimics the foraging behavior and movements of bird's flock, which keeps on trying to hunt their food sources. The said mechanism is employed to locate the solutions that solve the complex optimization problem. In PSO, randomly, a set of populations (i.e., particles or swarm) are generated and updated their position and velocity based on the information obtained from themselves. In PSO, each particle moves under certain velocity in their own position when flying to search their food source in multi-dimensional space. Optimal zones are determined, viz. heuristic search approach with the best experience of the individual particle (Pbest) or whole swarms (Gbest) to modify position toward global fitness (i.e., food source).

In PSO, the cognitive and social parts represent the rate of change of velocity of the particles based on self-fly experience and neighborhood particle experience. In the present work, an evolutionary operator (i.e., mutation to enhance the diversity in search space) is introduced to maintain numerous non-dominated solutions to store in the external archive. PSO differ from multi-objective particle swarm optimization-based crowding distance (MOPSO-CD) method particularly in the selection of cognitive and social leader by using Pareto dominant and crowding distance approach. MOPSO-CD parameters (mutation, inertia weight, swarm size, iterations) are sensitive toward solution accuracy (i.e., local or global minima) and convergence rate [18]. High inertia weight tends to facilitate initially toward global exploration and low inertia weight conducts a localized search as a result of poor exploitation [19]. Large population size could generate multiple global or local solutions [20], whereas the solution accuracy might not result in global solution always with small population size. PSO might not yield global fitness in one iteration, because the particles survive is intact with one iteration, corresponding to the next. Furthermore, an individual particle can finally move toward global while conducting a heuristic search in a multi-dimensional space provided, and they have initialized with a maximum number of iterations [19]. A large number of generations (i.e., iterations) increase the likelihood in locating the global fitness solution, but the amount of gain in solution accuracy must compensate with the computation or processing time and efforts spent. Note that, till date, no universal approach defined yet in selecting the appropriate choice of parameters of PSO. In the present work, PSO parameters are optimized by conducting a systematic study with a goal of maximizing the fitness value. The conflicting objective functions (LB for MRR and SB for SR) are formulated with a simple mathematical equation to form a single response function for maximization (refer Eq. 4.7). Desirability function approach (DFA) is employed to carry out the said task. Note that the overall desirability (D_o) value found to vary in the ranges between zero and one. The D_o value close to one depicts the ideal value, whereas nearer to zero determines completely undesirable for optimization. The computation of global desirability value assigned as fitness function value for optimization of conflicting behavior of responses is done according to Eq. (4.7).

$$\text{Fitness (or) } D_o = \sqrt[2]{\left(\frac{\text{MRR} - \text{MRR}_{\min}}{\text{MRR}_{\max} - \text{MRR}_{\min}}\right)^{w_1} \times \left(\frac{\text{SR}_{\max} - \text{SR}}{\text{SR}_{\max} - \text{SR}_{\min}}\right)^{w_2}}$$

(4.7)

where $\text{SR}_{\max}$, $\text{SR}_{\min}$, $\text{MRR}_{\max}$, and $\text{MRR}_{\min}$ are the corresponding maximum and minimum values of SR and MRR, respectively.

To optimize AWJM process considering simultaneously maximizing the production rate (industrial perspective of economical machining) and minimizing surface roughness (customer perspective for proper functioning during service life) is a tedious task. This occurs due to the complex nature of responses with the inputs. This results in many optimal solutions, which sometimes lead to sub-optimal solutions. Selecting one solution from multiple solutions is difficult, and this problem requires the study of a few case studies. Three cases are studied as follows: (a) equal importance (weights, w) to both objective functions ($W_1 = W_2 = 0.5$), (b) assigning maximum importance to MRR ($W_1 = 0.9$, and $W_2 = 0.1$), and (c) assigning

Table 4.7 MOPSO-CD parameters and operating levels

Variables and its ranges	Best value
Mutation probability (0.1–0.3)	0.18
Inertia weight (0.1–1.0)	0.8
Swarm size (10–100)	40
Maximum generations (10–100)	80

maximum importance to SR ($W_2 = 0.9$, and $W_1 = 0.1$). The highest fitness value obtained from the studied three cases is recommended as an optimal condition for getting better quality characteristics in AWJM process.

MOPSO-CD parameters (i.e., inertia weight, mutation probability, swarm size, and generations) are sensitive to solution accuracy and computation time. Improper choice of said parameters might trap at local minima solutions. Tuning of parameters poses a greater probability to hit global minima [21–23]. A systematic study was performed on the algorithm parameters when varied within their respective levels (refer Table 4.7), and recorded their fitness values (refer Fig. 4.7). The highest fitness (optimized parameter) value corresponds to each parameter thus selected to avoid

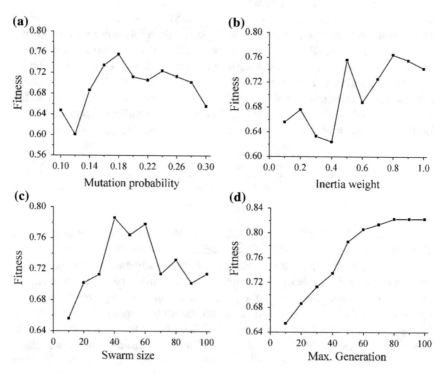

Fig. 4.7 MOPSO-CD parameter study of fitness versus **a** mutation probability, **b** inertia weight, **c** swarm size, and **d** maximum generations

Table 4.8 AWJM optimized conditions for different case studies

Case studies	Fitness value	Control variables (A, B, and C)	Responses (MRR and SR)
Case 1: W_1 (MRR) = W_2 (SR) = 0.5	0.8227	1.5, 150, 100	656.03 mm^3/min and 0.234 μm
Case 2: W_1 (MRR) = 0.9, and W_2 (SR) = 0.1	0.8932	3.1, 150, 100	866.67 mm^3/min and 0.348 μm
Case 3: W_1 (MRR) = 0.1, and W_2 (SR) = 0.9	0.8792	1.5, 116, 100	345.72 mm^3/min and 0.194 μm

local minima solutions (refer Fig. 4.7). The resulted optimized parameters of particle swarm optimization are presented in Table 4.7.

The MOPSO-CD determined optimal abrasive water jet machining conditions after fine-tuning of algorithmic parameters. The results of optimal conditions studied for three different cases are presented in Table 4.8. It is to be noted that the highest fitness value was obtained corresponds to case 2 (i.e., maximum importance assigned for MRR). Therefore, the set of abrasive water jet machining parameters corresponds to case 2 is recommended as the optimal set to yield better MRR and SR.

4.3.5 Confirmation Experiments

After the optimization, work is also carried out for the confirmatory analysis/experimentation to confirm the results obtained via. MOPSO-CD method. The confirmatory tests are performed based on the optimal setting obtained using MOPSO-CD method and the confirmatory results are tabulated in Table 4.9. Important to note that the optimal levels recommended by MOPSO-CD method are not among the combination of L_{27} orthogonal array experiments of Table 4.3. This occurs due to the multifactor nature of Taguchi experimental design (i.e., $3^5 = 243$). It

Table 4.9 Confirmatory results

Case studies	Control variables (A, B, and C)	Responses (MRR and SR) via MOPSO-CD method	Responses (MRR and SR) via confirmatory experiments
Case 1: W_1 (MRR) = W_2 (SR) = 0.5	1.5, 150, 100	656.03 mm^3/min and 0.234 μm	654.12 mm^3/min and 0.203 μm
Case 2: W_1 (MRR) = 0.9, and W_2 (SR) = 0.1	3.1, 150, 100	866.67 mm^3/min and 0.348 μm	864.12 mm^3/min and 0.355 μm
Case 3: W_1 (MRR) = 0.1, and W_2 (SR) = 0.9	1.5, 116, 100	345.72 mm^3/min and 0.194 μm	347.32 mm^3/min and 0.190 μm

is also observed that optimal values obtained via confirmatory experiments found acceptable and satisfactory with that of the experimental results. The results of the prediction performance of an optimization tool are in good agreement with less than 10% deviation with the experimental material removal rate and surface roughness.

4.3.6 SEM Study of Machined Surfaces

Additionally, machined surface of zirconia (ZrO_2) composite is analyzed using scanning electron microscopic (SEM). The model ZEISS EVO-Series Scanning Electron Microscope EVO 18 manufactured by ZIESS is used for the study. The machined surface obtained at optimal setting using MOPSO-CD method, i.e., case 3, $A = 1.5$ mm, $B = 116$ MPa, and $C = 100$ mm/min, is considered for the SEM image. The scanning electron microscopic images are shown in Fig. 4.8.

It is seen from Fig. 4.8a–c that, there are some alternations such as abrasive grain marks and clustered zirconia (ZrO_2) particles spread on the machined surfaces. This shows that, metal removal is by the impact of the abrasive particles with water pressure. Some of the places, hole patch is shown (Fig. 4.8b), indicate the erosion of ceramic materials with sharp corners of abrasive particles. This is also because of the lower cutting speed (C) and a higher standoff distance (A). In addition, a cluster

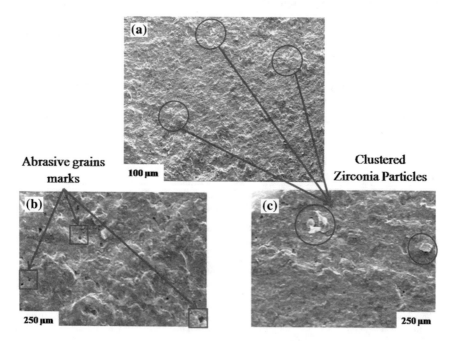

Fig. 4.8 SEM images of zirconia (ZrO_2) composite in AWJM process at ($A = 1.5$ mm, $B = 116$ MPa, and $C = 100$ mm/min

of zirconia particles (Fig. 4.8c) is fragmented on one side on the machined surface. This is due to brittle fracture and uneven erosion of workpiece surface results. Also, due to the lesser working pressure (B), a larger stand of distance (A) and larger nozzle speed (C). So, in order to get the optimal, lower values of A, B and C are considered as optimal during the machining of zirconia (ZrO_2) composites in AWJM, which directly gives better and smooth surfaces as well as higher MRR, produce lesser environmental issues during machining in AWJM.

4.4 Summary

This chapter presents the machining performance of green machining process on ceramic composites using abrasive water jet machining (AWJM), a commonly known as green machining process. Taguchi L_{27} orthogonal array is used for experimentation considering the three independent parameters like a standoff distance (A), working pressure (B), and nozzle speed (C). The parameter like MRR and SR is considered as a response or output parameter in this study. Based on the experimental results, parametric analysis, regression analysis, mathematical model, and optimization following conclusions are drawn,

- Green machining process, i.e., AWJM process is capable and adequate in the machining of zirconia (ZrO_2) composites.
- From the ANOVA: parameter standoff distance (A) followed by work pressure (B) showed dominant effect for SR and the optimal setting, i.e., $A_1B_2C_2$ combinations resulted from optimal SR. Similarly, work pressure (B) showed dominant effect on MRR compared to the other and combination $A_3B_3C_2$ resulted from optimal MRR.
- From the optimization: MOPSO-CD provides most optimal results for green machining process and the optimal setting obtained as to standoff distance (A) = 1.5 mm, working pressure (B) = 116 MPa, and nozzle speed (C) = 100 mm/min. The optimal values facilitate to yield better surface quality, higher MRR, and improved productivity.
- The optimal setting provides optimal responses such as the higher MRR, and better SR, which have less influence on a generation of environmental issues aroused during the machining of zirconia (ZrO_2) composites in AWJM process.
- Additionally, prediction models are developed for MRR and SR for optimal prediction of AWJM responses. The result shows that predicted responses are close and satisfactory with experimental results.
- From the SEM images, machined surface of zirconia (ZrO_2), composite found the smooth and uniform distribution of surface during machining in AWJM process.
- At last, a confirmatory test is performed to verify the experimental results. The result shows that confirmatory results and experimental results are comparable.

Finally, it is concluded that the AWJM process is adequate for machining of zirconia (ZrO_2) composites under green machining environment. The parametric setting obtained from the analysis can be used as the optimal setting for the AWJM process during the machining of zirconia (ZrO_2) composites. Also, the developed empirical models for AWJM responses can be employed for prediction and optimization of response parameters in manufacturing industries.

References

1. T.B. Thoe, D.K. Aspinwall, M.L.H. Wise, Review on ultrasonic machining. Int. J. Mach. Tools Manuf. **38**(4), 239–255 (1998)
2. I. Denry, J.R. Kelly, State of the art of zirconia for dental applications. Dent. Mater. **24**(3), 299–307 (2008)
3. G.S. Choi, G.H. Choi, Process analysis and monitoring in abrasive water jet machining of alumina ceramics. Int. J. Mach. Tools Manuf. **37**(3), 295–307 (1997)
4. E. Siores, W.C.K. Wong, L. Chen, J.G. Wager, Enhancing abrasive waterjet cutting of ceramics by head oscillation techniques. Ann. CIRP **45**(1), 327–330 (1996)
5. J. Holmberg, J. Berglund, A. Wretland, T. Beno, Evaluation of surface integrity after high energy machining with EDM, laser beam machining and abrasive water jet machining of alloy 718. Int. J. Adv. Manuf. Technol. **100**(5–8), 1575–1591 (2018)
6. E.O. Ezugwu, Z.M. Wang, A.R. Machado, The machinability of nickel-based alloys: a review. J. Mater. Process. Technol. **86**(1–3), 1–16 (1999)
7. D.K.M. Tan, A model for the surface finish in abrasive-waterjet cutting, in *8th International Symposium on Jet Cutting Technology* (1986), pp. 309–313
8. R.M. Samson, T. Geethapriyan, A.A. Raj, A. Ashok, A. Rajesh, Parametric optimization of abrasive water jet machining of beryllium copper using Taguchi grey relational analysis, in *Advances in Manufacturing Processes* (Springer, Singapore, 2019), pp. 501–520
9. Jagadish, S. Bhowmik, A. Ray, Prediction and optimization of process parameters of green composites in AWJM process using response surface methodology. Int. J. Adv. Manuf. Technol. **87**(5–8), 1359–1370 (2016)
10. M.M. Korat, G.D. Acharya, A review on current research and development in abrasive waterjet machining. J. Eng. Res. **4**, 423–432 (2014)
11. J. Wang, Machinability study of polymer matrix composites using abrasive waterjet cutting technology. J. Mater. Process. Technol. **94**, 30–35 (1999)
12. W. Konig, S. Rummenholler, Technological and industrial safety aspects in milling FRP. ASME Mach. Adv. Comp. **45**(66), 1–14 (1993)
13. S. Kalirasu, N. Rajini, J.T.W. Jappes, Mechanical and machining performance of glass and coconut sheath fibre polyester composites using AWJM. J. Reinf. Plast. Compos. **34**(7), 564–580 (2015)
14. J. Kechagias, G. Petropoulos, N. Vaxevanidis, Application of Taguchi design for quality characterization of abrasive water jet machining of TRIP sheet steels. Int. J. Adv. Manuf. Technol. **62**(5–8), 635–643 (2012)
15. P.J. Ross, Taguchi Techniques for Quality Engineering: Loss Function, Orthogonal Experiments, Parameter and Tolerance Design. No. TS156 R12 (McGraw-Hill, New York, 1988)
16. G.C.M. Patel, P. Krishna, M.B. Parappagoudar, Squeeze casting process modeling by a conventional statistical regression analysis approach. Appl. Math. Model. **40**(15–16), 6869–6888 (2016)
17. R. Eberhart, J. Kennedy, A new optimizer using particle swarm theory, in *MHS'95, Proceedings of the Sixth International Symposium on Micro Machine and Human Science, 1995* (IEEE, 1995)
18. F. Van den Bergh, A.P. Engelbrecht, A study of particle swarm optimization particle trajectories. Inf. Sci. **176**(8), 937–971 (2006)

19. R.C. Eberhart, Y. Shi, Comparison between genetic algorithms and particle swarm optimization, in *International Conference on Evolutionary Programming* (Springer, Berlin, Heidelberg, 1998)
20. G. Zhang, Y. Li, Y. Shi, Distributed learning particle swarm optimizer for global optimization of multimodal problems. Front. Comput. Sci. **12**(1), 122–134 (2018)
21. G.C.M. Patel, P. Krishna, M.B. Parappagoudar, P.R. Vundavilli, Multi-objective optimization of squeeze casting process using evolutionary algorithms. Int. J. Swarm Intell. Res. (IJSIR) **7**(1), 55–74 (2016)
22. G.R. Chate, G.C.M. Patel, A.S. Deshpande, M.B. Parappagoudar, Modeling and optimization of furan molding sand system using design of experiments and particle swarm optimization. Proc. IMechE Part E J. Process Mech. Eng. **232**(5), 579–598 (2018)
23. G.C.M. Patel, P. Krishna, M.B. Parappagoudar, P.R. Vundavilli, S.B. Bhushan, Squeeze casting parameter optimization using swarm intelligence and evolutionary algorithms, in *Critical Developments and Applications of Swarm Intelligence* (IGI Global, 2018), pp. 245–270

Index

A
Analysis of Variance (ANOVA), 14, 22–25, 29, 30, 34, 40, 42, 48, 62–64, 69

C
Ceramic, 1, 3, 9, 33, 51, 52, 55, 68, 69
Composite, 3, 9, 14, 33–36, 38–40, 43, 45, 47, 48, 52, 53, 68–70

D
Difficult-to-cut, 1, 8, 14, 51

G
Green, 7, 9, 10, 14, 16–20, 22, 27–30, 33–36, 43, 45, 47, 48, 52, 54, 56, 69

M
Material removal rate, 6, 9, 14, 17–25, 27–30, 34, 37, 38, 41–45, 47, 48, 52, 55–60, 62–65, 67–69

O
Optimization, 1, 9, 23, 24, 26, 27, 29, 30, 43, 47, 48, 54, 55, 64, 65, 67–70

P
Polymer, 1, 9, 14, 33, 34, 47

R
Regression, 14, 22, 23, 30, 34, 41, 42, 47, 62, 69

S
Silicon carbide, 1, 5, 9
Surface roughness, 5–9, 14, 21, 23, 24, 33, 34, 38, 39, 52, 57, 59, 61, 63, 65, 68
Sustainability, 10, 33

Z
Zirconia, 52, 53, 68–70

Printed in the United States
By Bookmasters